The Best of Technology Writing 2008

DIGITALCULTUreBOOKS is a collaborative imprint of the University of Michigan Press and the University of Michigan Library.

Clive Thompson, Editor

The Best of Technology Writing 2008

THE UNIVERSITY OF MICHIGAN PRESS AND
THE UNIVERSITY OF MICHIGAN LIBRARY
Ann Arbor

Copyright © by the University of Michigan 2008
All rights reserved
Published in the United States of America by
The University of Michigan Press and
The University of Michigan Library
Manufactured in the United States of America
⊗ Printed on acid-free paper

2011   2010   2009   2008   4   3   2   1

A CIP catalog record for this book is available from the British Library.

ISBN-13: 978-0-472-03327-0
ISBN-10: 0-472-03327-1

ISSN 1938-7113

Contents

Introduction

The scientist Norbert Wiener used to say that "a society is defined by its technique."

It's a weird aphorism, at first blush. But when Wiener said "technique," he was riffing off the Greek root *tekhne*—which means "skill" and which is also the root of the word *technology.*

What Wiener meant is really pretty simple: New technologies endow us with new skills, and those skills define how a society operates. Back in 3300 BC, tinkerers figured out how to smelt bronze, and suddenly they could produce metals with then-unheard-of-strength—metals which in turn produced drastically deadlier weapons and radically more efficient farming tools. A few thousand years later, the telegraph wire allowed people to send a message from Europe to the United States instantaneously, and suddenly the pace of business and news sped up to a degree that seemed almost insane. (Previously, a message took *two weeks* to cross the ocean on a steamer.) Twenty years ago, computer scientists figured out a technique for compressing a pop song into an e-mailable three-meg file, and, *boom:* they reshaped the entire recording industry, dooming the slow-to-react labels and empowering amateurs to reach the globe from their bedrooms. They probably didn't intend these things to happen, but they did. Every new technology changes society. Usually the changes are small and meaningless. And then, every once in a while, they're huge, weird, and totally unexpected.

What I love about good technology writing is that it captures these changes. Often it's not really about technology at all, but about people. *What happens to us when we're given strange new powers?*

That question is at the heart of every article in this anthology. Sure, the writers collected here are all technically savvy; they're superb at describing, in suitably nongeeky lay terms, how the gizmo works. But they're even better at teasing out the odd—and sometimes scary—transformations new technology wreaks upon the world.

Sometimes the changes are merely delightful, as when inventor Dave Arnold rejiggers well-known foods to produce a new-age corn dog and a superdistilled, "breathable" gin and tonic (a nifty feat of modern alchemy described by Ted Allen in "Doctor Delicious"). Or sometimes the changes are politically subversive, as when Robin Mejia describes how a young human-rights activist uses satellite images to document genocide, peering from the sky in order to bypass the control of local dictators. More often the social transformations are subtle and unheralded. A lot of everyday technologies creep into our lives, very slowly, and it takes a superb writer to stand back and point out what's happened without our noticing. I'm thinking here of "Say Everything," Emily Nussbaum's description of how young people, growing up in an age of omnipresent Facebooking, blogging, and Flickr-picture-posting, have embraced a radically looser sense of personal privacy than their parents—a shift that Nussbaum credits with having created the biggest cultural generation gap since rock and roll. (Speaking of personal privacy, I should disclose that Nussbaum is my wife, though her story is so good I decided to risk the charge of marital logrolling by including it.)

Technology writing is a sneakily broad format—a kind of catchall. Because technological change influences every aspect of our lives, tech writers get to cover everything—architecture, design, health care, law, sports—so long as there's a good gearhead angle. Charles Graeber's story about the frantic attempt of Cannonball Run freaks to drive across the United States in barely 32 hours is filled with hilarious, outrageous, cop-radar-defeating detail. And yet, it also works, on another level, as a rollicking sports story about crazy daredevils. John Seabrook's superb article "Fragmentary Knowledge" is a straightforward piece about archeology, which describes how rival teams of archeologists tried to figure out what the Antikythera Mechanism—a mysterious artifact from the first century BC—actually is. But since the story involves the use of a newfangled eight-ton X-ray machine, and since the Mechanism *might* be the world's first computer, it's also a technology story with a gorgeously meta twist: cutting-edge tools being deployed to investigate the origins of cutting-edge tools. And then again, "Fragmentary Knowledge" is also a tale of people striving for knowledge and bragging rights to a great discovery—and *that* human drama is as old as the hills. Technology changes how we do things and what we do, but it doesn't change human nature: It just amplifies it.

There is a ton of science—really marvelous science—in these pages. Because technology and science have such a symbiotic relationship, technology writing has always been a sort of stealth form of science journalism. It's often "practical" people—inventors, farmers, doctors—who stumble upon a new scientific principle while trying to create some gee-whiz new device. And so it goes in these pages too. Over and over again, we start off reading about a new gizmo and wind up hip deep in some fascinating and breakthrough science. In "The Brain on the Stand," Jeffrey Rosen describes lawyers who are trying to exculpate their clients by using "brain-scanning" machines to show that these accused murderers' brains do not possess crucial details of a crime scene. Of course, this technique plunges us into some gripping and freaky questions about the mind: *What do memories actually mean? How does the brain record them?* Not to mention some even more unsettling questions about the future of human rights: *Can a judge compel you to have your brain scanned, against your will—or would that violate your Fifth Amendment protection against self-incrimination? Can your brain testify against you?* No one yet knows the answers to these questions, which is precisely why I'm glad we have so many good technology writers around to ask them.

Indeed, you could say that we're living in a golden age of technology journalism. Ten or 15 years ago, tech writing was confined to small media ghettos. There were a bunch of terrific fan magazines, like *Byte* magazine and its now countless progeny, but nobody other than nerds like me read them. When mainstream newspapers and magazines covered technology, it was mostly gadget reviews—little 600-word, *Consumer Report*-esque assessments of the latest gewgaw, with a one- to five-star rating. At best, you could find some tolerably okay writing in the business pages, where eyeshade-sporting investors flocked looking for some new breakthrough. But, otherwise, mainstream media devoted precious little attention to technology. The front page was reserved for news with more traditional "importance," like, say, partisan politics.

Today, however, things are radically different. The top newspapers and magazines are stumbling all over themselves to cover technology—putting it not just on the front but in the culture and Op-Ed pages too. Why? You can mostly thank the internet boom, which made the cultural and political impact of technology completely unignorable. Everything's been hit. Ebay and Amazon and a zillion tiny mom-and-pop shops have gamed the economy into something uncannily different from what it was before. The cab drivers who ferry me around Manhattan spend their entire shift chitchatting with their families in India and Pakistan via dirt-cheap voice-over-IP phone calls that are almost too inexpensive to meter. And even electoral politics are bending under the gale force wind of the net, with YouTube gaffe clips destroying candidates and online fund-raising vaporizing the influence of the wealthy Republican and Democratic donors who once drove their parties like personal go-karts. No wonder the nation's editors are in a lather to cover this stuff, and cover it well.

One thing you'll notice about this anthology is that I've disproportionately populated it with stories from a few big, glossy magazines like *Wired, Popular Science,* the *New Yorker,* and the *New York Times Magazine.* Obviously, technology writing takes place everywhere now. But I'll admit I have a bias: I am a devoted fan of long-form journalism—stories with a documentarian's view, stories that give us scenes and characters and rummage deeply in the implications of a given technology. This sort of work needs time and space, and only a small number of magazines still have the resources to provide them.

I'd originally hoped to include some writing from blogs. As it happens, I spend a large part of my day reading Web sites that regularly produce superb, nuanced takes on the social implications of technology. But when I tried to pick some excerpts, I realized that online writing often takes place as a dialogue: Author A writes a short post that writer B comments on, prompting A to write her own reply, while bloggers C through L are weighing in too. And, in the middle of it all, *presto:* Blogger D will pen a stunningly brilliant, well-reasoned point in gorgeous prose—something that I'd love to put in an anthology such as this one. But what I discovered is that, if you try to excerpt D's writing, it simply won't make sense because it's entwined so talmudically with everyone else's. This is itself, of course, an example of cultural change driven by technology: Before the net, it wasn't possible for people worldwide to peer at each other's thoughts and then fire back a rapid retort.

Maybe 10 years from now this book will be a sort of hybrid: a digital artifact you hold in your hands that includes not just the sort of lengthy, stand-alone think pieces I'm offering you now but also a curated collection of the salonlike conversations going on online. And maybe it'll update itself in real time: You'll open it to discover that the book has grown another chapter or that one of the authors has added some new thoughts or that someone halfway around the world has inserted a really astute commentary. What exactly would you call such a thing? Will it still be regarded as a book? And will it be a sign of cultural decline? Do we need the permanence, the unrevisability, of paper to freeze ideas in place so we can deeply imbibe them? Or will this new genre of device spur some subtly new way of reading, the kind of newfangled practice that will cause future parents to complain that they don't understand their kids' books anymore?

Sounds like a good technology story. Maybe someday we'll be writing it.

Doctor Delicious

*When the world's best chefs want something that
defies the laws of physics, they come to one man:
Dave Arnold, the DIY guru of high-tech cooking.*

Dave Arnold would like to fix you a gin and tonic. Sound
good? It will be. It will be very, very good. It will be like no
gin and tonic you have ever seen or tasted in your life. It will
also be considerably more involved, shall we say, than cracking open the Tanqueray and Schweppes.

First, Arnold believes, he must clarify the lime juice.
Why? Because his uncompromising conception of culinary
perfection requires that gin and tonics be completely crystalline clear, that's why. And so, from a closet in the back of
a teaching kitchen at the French Culinary Institute (FCI) in
New York City, behind a door labeled "Caution: Nitrous
Oxide in Use," Arnold wheels out a cart piled high with laboratory equipment—a rotary evaporator (rotovap) that he
salvaged from Eli Lilly on eBay, cheap, and that he has
jerry-rigged for just this sort of thing. At his side, FCI chef
and VP Nils Noren supports a somewhat wobbly condenser
as Arnold pours a liter of freshly squeezed lime juice, pale
green and cloudy with pulp, into a teardrop-shaped Pyrex
vessel. Because heat would destroy the flavors and aromas of

the elixir, Arnold brings the vessel just above room temperature by partially submerging it in a bath of precisely regulated warm water. He then connects it to a vacuum so that the juice will vaporize at low temperatures.

Arnold flips the switch. The machine gurgles and hums, the vessel spins merrily, the lime vapor drifts up into the condenser, and an absolutely clear liquid begins dripping into a beaker. The result smells like lime, but it's lost much of its punchy flavor in distillation. So Arnold works to bring his clarified juice back into balance. From a series of plastic bottles, he adds 4.5 percent powdered citric acid, 1.5 percent malic acid, and 0.1 percent succinic acid to the solution; places the beaker atop an electromagnetic stirrer; drops in a little Teflon-coated magnetic bar; and flips the switch. Instantly, the bar begins spinning, whipping up the liquid and dissolving the powders. Voilà! Clearlime, Arnold calls it. A touch of quinine powder and some simple syrup (two to one, sugar and water), some water, and, after a couple hours of labor, he's halfway there.

Now he custom makes his own "gin," really just a neutral spirit infused with whatever aromatics are catching Arnold's fancy and then distilled (the latter part of which is, in fact, illegal—but hey, it's all in the name of science). Today it will be two cucumbers, celery ribs, roasted orange slices, and one bunch each of cilantro and Thai basil, all coarsely chopped and added to a fifth of Absolut vodka. Everything goes into the vessel and back on board the rotovap, and another beaker is filled.

The two liquids are combined about one to one, heavily carbonated with a healthy injection of CO_2 (Arnold loves carbonation), and chilled for 20 minutes to a blistering cold in a freezer (he hates it when ice melts in his drinks). And so, sans rocks, sans garnish, Arnold pours the concoction into champagne flutes and serves it.

"I like my drinks stiff," he notes, and he is not kidding. This take on the G&T is, literally and figuratively, a distillation of the classic's flavors. It's a pure, Platonic ideal of the G&T, strong as a martini. The sensation is not so much of drinking something as it is of breathing it, the effervescence unusually intense and refreshing, the flavors and aromas magnified, permeating the palate and nose with a sharp, aggressive, limey crispness, underscored with soft notes of cilantro, roasted orange, and cuke. And it only took three hours.

"It's a crazy level of things you have to do to get the product I want," Arnold says, "but here's what happens when you do everything possible to get something the way you want it. Yeah, sure, it's ridiculous, but . . ."

You should see how he cooks a steak.

BIGGER MOTORS

Dave Arnold is the man behind the curtain of today's hottest movement in cooking, molecular gastronomy. He's the Q to James Bond as embodied by esteemed mad-scientist chef Wylie Dufresne. A former paralegal, performance artist, and, briefly, Domino's Pizza driver, Arnold has become the go-to gearhead for machines and techniques to help chefs realize their wildest culinary fantasies. And wild they are: Carbonated watermelon. Gelatin spheres with liquid centers that pop in your mouth. Broths and sauces whipped into foams. Shrimp flesh extruded into "noodles." Hot-center desserts with exteriors flash frozen by liquid nitrogen. Vanilla beans sizzled table side with lasers. (It should be noted that Arnold disapproves of sizzling things table side with lasers, because of safety concerns—which, for reasons that will soon become clear, is funny.)

All those culinary pyrotechnics can't happen without a

lot of R&D. That's Arnold's specialty. The 36-year-old, salt-and-pepper-haired, wildly enthusiastic food lover is part artist, part scientist, part self-taught machinist, and, of course, exuberant cook. Armed with a BA in philosophy from Yale and an MFA from Columbia but largely self-taught in the areas of cooking and engineering, he was hired at FCI in 2005 as director of culinary technology, a new department augmenting the school's traditional instruction with scientific techniques, tools, and rigor. He instantly became one of the most popular instructors there.

Perhaps that popularity has something to do with his unbridled excitement at the power of technology to create deliciousness. Take, for example, how he goes about improving the immersion blender—the handheld blender "stick" that allows cooks to puree foods in sauce pots and bowls. For Arnold's purposes, the blenders on the market are far too weak, so he rigged one together using an 18-volt battery and the motor from a DeWalt cordless drill, resulting in a stick blender as strong as a commercial milkshake machine. "Just unbelievably powerful," he says. "I get such a huge vortex, I can make stuff as smooth as you can in a Vita-Mix."

Or consider his take on the humble corn dog. "The problem with them is, one, you don't get that high-heat, cooked flavor in the sausage, and two, the batter is never cooked right next to the sausage." His vision, inspired by a classic German cake called *baumkuchen,* which is baked in layers on a rotating spit: Skewer the dogs on a rotisserie, get a little char on them, and then apply batter in thin coats so that each one is perfectly cooked.

It's this kind of ingenuity that has propelled him into the kitchens of the most celebrated chefs cooking today. On a given afternoon, he could be showing David Chang how to carbonate sake at one of Chang's Momofuku restaurants in

New York or creating a syringe for Johnny Iuzzini, the pastry chef at Jean-Georges, also in New York, to layer a hot flavored gelatin atop a cold one for a modern take on the pousse-café. Or ripping apart his espresso machine and modifying it to mimic a hand-pulled shot. "He's nothing shy of a genius," says chef Charlie Trotter, of the legendary Chicago restaurant that bears his name, who met Arnold at a fusion-cooking conference in Madrid last year. "He's helping chefs take their food to the next level."

TRY ANYTHING

Poised with a lance and wearing a welding jacket, his wife at the ready with her camera, Dave Arnold is preparing to face off with a dragon. Actually, with a snowblower. A snowblower that he has mounted on a tripod and rigged to spray flaming kerosene vapor. At himself.

This is during art school, you'll understand.

"The idea was that if I could jam the lance into where the blower was going, I could stop the blower, and I would win," he explains. Instead, the dragon won, and Arnold was engulfed in flames. "I learned that what happens when you catch on fire is you don't 'stop, drop, and roll,'" Arnold says. "You start running around to try to get away from yourself. Luckily, I had a bunch of friends there who tackled me. I ended up having to go to the hospital."

Arnold's typical projects, though no less extreme, aren't always quite so hazardous. Harold McGee, author of the seminal 1984 classic on the science of the kitchen, *On Food and Cooking,* recalls a long day spent with Arnold trolling exotic markets all over Manhattan, solely because Arnold insisted that McGee experience an ingredient he had just discovered: giant water bug essence from Thailand. "It smells like a combination of really strong pear aroma with a little bit of nail polish in the background," McGee says. "He just wants to find everything and experience everything."

That kind of fearless curiosity came early. Growing up as an only child (until the age of 15) in the New York area, Arnold says, "I ate everything." He also took up culinary experimentation early. Aside from his childhood specialty, chicken cooked in parchment with his own proprietary spice mix, he was the self-styled "breakfast king," getting up early on weekends to make breakfast in bed for his parents. Among his more ambitious adventures: deep-fried beignets. "Looking back," he notes, "I don't think fifth graders should deep-fry by themselves while their parents are asleep."

Arnold has tech in his genes: His mother is a doctor and his father an engineer, as were both of his grandfathers. He had always imagined an academic career in science. But at Yale, he went with liberal arts course work, attributing the decision to boredom and "a little bit of ADD." As a junior, he started dating the woman who is now his wife—Jennifer Carpenter, then an architecture student interning with Cesar Pelli—and thought it might be smart to dabble in course work related to her field, "because then I would have something else to talk to her about." So he signed up for a sculpting class. "They taught me how to weld, and I was like, 'This is amazing.' I was like, 'What? I can make big things from metal that move and spin?'"

He fell in love with building machinery. He also decided to go to art school. While at Columbia, food occasionally found its way into his work—one performance piece he contemplated was fashioning a model of the city Nagasaki from gingerbread and blowing it to pieces.

As it happens, the work with arc welders and flaming snowblowers proved to be useful training. In the late 1990s, he and Carpenter moved into an illegal loft on 38th Street that lacked a kitchen. Using a dorm fridge, a hotplate, and a utility sink from Home Depot, he created a roll-away kitchen that could be hidden in case the landlord came sniffing around. When Arnold and Carpenter noticed that the landlord never actually did come around, they became emboldened—and Arnold discovered restaurant equipment auctions.

"The first thing I bought was a double-glass sliding-door deli case" for $65, he recalls. "That thing changed my life. You could see all the food in it. I had a party once—and this was before I had a soft-serve machine—and I had something like eight cases of soft-serve, five cases of beer, three cases of champagne, a ham, a turkey, and all the noshes for everything, and the thing wasn't even full." Soon thereafter, he bought a four-gallon commercial deep fryer from a shuttered Mexican joint in the financial district and rolled it home on a hand truck—in the snow. Then he got a commercial broiler, known in the trade as a salamander. Then a convection oven. He also began customizing his equipment, starting when the salvaged convection oven didn't perform to his liking.

It was around this time that he discovered wd-50, Wylie Dufresne's acclaimed experimental restaurant on the Lower East Side of Manhattan. Arnold quickly became a regular. He asked Dufresne for a kitchen tour; the two hit it off; and, before long, they became friends. (It didn't hurt that Dufresne had become interested in Arnold's sister-in-law Maile Carpenter, who he had met in her capacity then as *Time Out New York*'s food editor; the two are now engaged.)

"He was the one who said, 'You can take your tech and machine knowledge and your cooking knowledge and bring them together,'" Arnold says. Dufresne was (and still is) interested in sous vide and other methods of cooking food slowly in liquids, at low temperatures, until the exact moment it is done. Early in his work with Arnold, he complained that doing so with traditional equipment was too difficult. Indeed, it is virtually impossible to keep water at a constant, very low temperature for hours and hours on a stovetop. Dufresne asked Arnold if he could find him an immersion circulator, a thermostated water bath common to the most rudimentary chemist's workshop. Arnold replied, "Well, I don't know what one is, but I guarantee I can get it." He took Dufresne's money and started scouring eBay. A collaboration was born.

I recently toured wd-50's kitchen to get a look at the arsenal of tools that Arnold has made or modified and that have become essential to Dufresne's cutting-edge cuisine. Observing that fish proteins coagulate at 125 to 135 degrees ("That's when the muscle begins to contract and squeeze out that white, milky stuff, and that's when fish begins to dry out"), Dufresne told Arnold that he wanted to cook fish very slowly in a moist environment until the precise moment it reached those temperatures, in a much lower-temperature environment than the 212 degrees necessary to create steam. Arnold took parts from a humidifier, which converts water to vapor with sonic pulses rather than heat, added a heating coil to produce the moderate temperatures Dufresne was after, and, in effect, built him what is now called a vapor oven years before they were widely available. "And so," Dufresne says, "we're able to cook a moister piece of fish."

Nearby, sous-chef Jeffrey Fisher is experimenting with a vacuum fryer, modified by Arnold with a condenser and hoses to remove water vapor. The vacuum permits liquids to boil at much lower temperatures, a property that Fisher is exploring to fry chips of apple, garlic, and potato in oil without browning them. "The goal is green things stay green, white fruit chips stay white, that sort of thing," Fisher says. Unfortunately, he notes, the fryer is still retaining too much moisture—you can see droplets condensing on the under-side of the clear lid—and, as such, the chips are coming out squishy. Arnold took one look at the problem and an-nounced that the lid should be dome shaped rather than flat, so that water droplets would slide to the edge rather than fall on the food.

Not long after he began tinkering with Dufresne's equipment, Arnold was working up a proposal for a food museum and writing stories for *Food Arts* magazine. Editor Michael Batterbury noticed his interest in technology and tapped Arnold to write equipment reviews. Then, two years ago, administrators at the French Culinary Institute (Dufresne's alma mater) decided to create a new department focused on molecular gastronomy and went looking for a department head. Batterbury recommended Arnold for the job. "You don't want a chef to do this position," Arnold says. "You want someone who can figure out what the chef really wants, talk to the science and tech people, and be the liaison between the two. That's what I do here."

MAGIC MEAT GLUE

Perhaps the most fun to be had with experimental cooking comes from the magic potions known as hydrocolloids, a class of ingredients familiar to anyone who's perused the la-bels of processed foods—cellulose, xanthan gum, agar, algi-nate, carrageenan, gelatin—but that, until recently (with the exception of gelatin), were not a fine cook's ingredients. Generally, hydrocolloids are used to thicken, gel, or stabilize liquids; they can also be used to great effect to change tex-ture, enabling a chef to produce a foam that won't collapse or, in Arnold's case, to make a "sponge cake" with methyl cellulose that can be shot from a compressed whipped-cream canister onto a plate without requiring baking.

Despite their negative associations with junk food, most hydrocolloids actually come from natural sources. Agar and carrageenan are derived from seaweed, gelatin from cow and pig bones, and pectin from citrus and apples. Some of these additives, such as agar, a common thickener in Thai cooking, have been around for centuries; others, like transglutaminase (known in the industry as "meat glue"), are newer and can be used for some pretty out-there stuff—attaching chicken skin to a piece of fish, say, or gluing a piece of skate wing to a slab of pork belly. In his appearance on *Iron Chef* in 2005, Dufresne used transglutaminase to bind pureed fish into "noodles," which were toothsome and delicious, not to mention clever. (Full disclosure: I served as a judge on that episode. I voted for Dufresne to win, but my colleagues overruled me and gave the nod to Mario Batali.)

"The problem," Arnold says, "is these [additives] have been used for decades to make products with a longer shelf life, to reduce the fat, to make something that you can freeze, to make something that ships farther, to make something that's cheaper. And these are all things that, in the end, reduce quality. Chefs have started looking at these ingredients as a quality enhancer, something to be proud of. Most of the top people are using these products because they make food better. Hardly any of them talk about it, because it sounds gross. There are a couple of people who, like Wylie, they talk about using these things because they love the products and they're trying to rehabilitate their image. So there's use of these products for economy, and there's use for effect, and these chefs are using it for effect."

"Sometimes it's just about learning," Dufresne says. "It's about understanding. That's why bringing traditional chefs together with scientists is infinitely interesting, because even if, at the very least, all they do is help explain things, and help us understand what's happening while we cook, then we're becoming better cooks."

When FCI hired Arnold, the plan was to build him a lab, which has yet to happen—he has his closet and a cubicle, and he scavenges most of his equipment used. ("I picked up a really good vacuum controller for cheap because some sucker listed it in the wrong category!") He's now in the market for a centrifuge, figuring that it will speed up the juice-clarification process, and he particularly dreams of getting a deal on a 3-D rapid-prototyping machine. Budget is an issue, not to mention the storage constraints of the two-bedroom apartment he shares with his wife and two young sons on the Lower East Side.

Down the road, Arnold is hatching plans to open the ultimate high-tech cocktail bar with pastry chef Iuzzini, focusing not on the retro, golden age drinks favored by most mixological temples but on an ultra-modern paradigm: still wines and juices carbonated to order with tongue-tingling intensity; rows of magnetic stirrers merrily whirling people's drinks in a chilling bath; rotovapping herbs and fruit for intense flavor; bourbon with soft, sweet nitrous-oxide bubbles; extremely cold drinks without the corruption of ice; super-chilled cocktail stirrers. "There are always new things you can do that are really delicious that no one is trying," Arnold says, "because they're so hyped up on getting back to some other place."

To his mind, this kind of problem solving isn't any sort of radical culinary departure. "People ask, 'Is this a fad?' I hope that the idea of trying to use everything at your disposal to make something better is never considered a fad, you know?" As Dufresne puts it, "I mean, an oven is technology. At one point, people were throwing sticks at animals and holding them over a spit, and that was a huge breakthrough."

McGee harkens back to Arnold's relentless quest for the perfect G&T. "He has this ideal of the french fry, the gin and tonic, so many things, and he's always trying to get to that ideal," he says. But ultimately, McGee ventures, he himself would probably prefer the old-fashioned kind: "For me, a gin and tonic is a tall drink that you sip. It's not a martini; it's a drink to quench your thirst. So I kind of like the standard one, with some Schweppes. I like the bursts of acidity from those little lime bits."

Of course, he knows, "if I were having this conversation with Dave, he would be saying, 'Well, if you like those little bursts of acidity, we can put some gelatin pearls in there, infused with Clearlime, so that whenever you bite one . . .'"

Say Everything

*As younger people reveal their private lives on the
internet, the older generation looks on with alarm
and misapprehension not seen since the early
days of rock and roll. The future belongs to the
uninhibited.*

"Yeah, I am naked on the internet," says Kitty Ostapowicz,
laughing. "But I've always said I wouldn't ever put up anything I wouldn't want my mother to see."

She hands me a Bud Lite. Kitty, 26, is a bartender at
Kabin in the East Village, and she is frankly adorable, with
bright-red hair, a button nose, and pretty features. She
knows it, too: Kitty tells me that she used to participate in
"ratings communities," like "nonuglies," where people
would post photos to be judged by strangers. She has a My-
Space page and a Livejournal. And she tells me that the internet brought her to New York, when a friend she met in a
chat room introduced her to his Web site, which linked to
his friends, one of whom was a photographer. Kitty posed
for that photographer in Buffalo, where she grew up, and
then followed him to New York. "Pretty much just wanted
a change," she says. "A drastic, drastic change."

Her Livejournal has gotten less personal over time, she

tells me. At first it was "just a lot of day-to-day bullshit, quizzes and stuff," but now she tries to "keep it concise to important events." When I ask her how she thinks she'll feel at 35, when her postings are a Google search away, she's okay with that. "I'll be proud!" she says. "It's a documentation of my youth, in a way. Even if it's just me, going back and Googling myself in 25 or 30 years. It's my self—what I used to be, what I used to do."

We settle up, and I go home to search for Kitty's profile. I'm expecting tame stuff: updates to friends, plus those blurry nudes. But, as it turns out, the photos we talked about (artistic shots of Kitty in bed or, in one picture, in a snow-drift, wearing stilettos) are the least revelatory thing I find. In posts tracing back to college, her story scrolls down my screen in raw and affecting detail: the death of her parents, her breakups, her insecurities, her ambitions. There are photos, but they are candid and unstylized, like a close-up of a tattoo of a butterfly, adjacent (explains the caption) to a bruise she got by bumping into the cash register. A recent entry encourages posters to share stories of sexual assault anonymously.

Some posts read like diary entries: "My period is way late, and I haven't been laid in months, so I don't know what the fuck is up." There are bar anecdotes: "I had a weird guy last night come into work and tell me all about how if I were in the South Bronx, I'd be raped if I were lucky. It was to-tally unprovoked, and he told me all about my stupid gener-ation and how he fought in Vietnam, and how today's navy and marines are a bunch of pussies." But the roughest mate-rial comes in her early posts, where she struggles with losing her parents. "I lost her four years ago today. A few hours ago to be precise," she writes. "What may well be the worst day of my life."

Talking to her the night before, I had liked Kitty: She

was warm and funny and humble, despite the "nonuglies" business. But reading her Livejournal, I feel thrown off. Some of it makes me wince. Much of it is witty and insightful. Mainly, I feel bizarrely protective of her, someone I've met once—she seems so exposed. And that feeling makes me feel very, very old.

Because the truth is, at 26, Kitty is herself an old lady, in internet terms. She left her teens several years before the revolution began in earnest: the forest of arms waving cell phone cameras at concerts, the MySpace pages blinking pink neon revelations, Xanga and Sconex and YouTube and Lastnightsparty.com and Flickr and Facebook and del.icio.us and Wikipedia and, especially, the ordinary, endless stream of daily documentation that is built into the life of anyone growing up today. You can see the evidence everywhere, from the rural 15-year-old who records videos for thousands of subscribers to the NYU students texting come-ons from beneath the bar. Even 9-year-olds have their own site, Club Penguin, to play games and plan parties. The change has rippled through pretty much every act of growing up. Go through your first big breakup and you may need to change your status on Facebook from "In a relationship" to "Single." Everyone will see it on your "feed," including your ex, and that's part of the point.

HEY NINETEEN

It's been a long time since there was a true generation gap, perhaps 50 years—you have to go back to the early years of rock and roll, when old people still talked about "jungle rhythms." Everything associated with that music and its greasy, shaggy culture felt baffling and divisive, from the crude slang to the dirty thoughts it was rumored to trigger in little girls. That musical divide has all but disappeared.

But in the past 10 years, a new set of values has sneaked in to take its place, erecting another barrier between young and old. And as it did in the '50s, the older generation has responded with a disgusted, dismissive squawk. It goes something like this:

> Kids today. They have no sense of shame. They have no sense of privacy. They are show-offs, fame whores, pornographic little loons who post their diaries, their phone numbers, their stupid poetry—for God's sake, their dirty photos!—online. They have virtual friends instead of real ones. They talk in illiterate instant messages. They are interested only in attention—and yet they have zero attention span, flitting like hummingbirds from one virtual stage to another.

"When it is more important to be seen than to be talented, it is hardly surprising that the less gifted among us are willing to fart our way into the spotlight," sneers Lakshmi Chaudhry in the current issue of the *Nation*. "Without any meaningful standard by which to measure our worth, we turn to the public eye for affirmation."

Clay Shirky, a 42-year-old professor of new media at NYU's Interactive Telecommunications Program, who has studied these phenomena since 1993, has a theory about that response. "Whenever young people are allowed to indulge in something old people are not allowed to, it makes us bitter. What did we have? The mall and the parking lot of the 7-Eleven? It sucked to grow up when we did! And we're mad about it now." People are always eager to believe that their behavior is a matter of morality, not chronology, Shirky argues. "You didn't behave like that because nobody gave you the option."

None of this is to suggest that older people aren't online, of course; they are, in huge numbers. It's just that it doesn't come naturally to them. "It is a constant surprise to those of us over a certain age, let's say 30, that large parts of our life can end up online," says Shirky. "But that's not a behavior anyone under 30 has had to unlearn." Despite his expertise, Shirky himself can feel the gulf growing between himself and his students, even in the past five years. "It used to be that we were all in this together. But now my job is not to de-mystify, but to get the students to see that it's strange or un-usual at all. Because they're soaking in it."

One night at Two Boots pizza, I meet some tourists vis-iting from Kansas City: Kent Gasaway, his daughter Han-nah, and two of her friends. The girls are 15. They have identical shiny hair and Ugg boots, and they answer my questions in a tangle of upspeak. Everyone has a Facebook, they tell me. Everyone used to have a Xanga ("So seventh grade!"). They got computers in third grade. Yes, they post party pictures. Yes, they use "away messages." When I ask them why they'd like to appear on a reality show, they ex-plain, "It's the fame and the—well, not the fame, just the whole, 'Oh, my God, weren't you on TV?'"

After a few minutes of this, I turn to Gasaway and ask if he has a Web page. He seems baffled by the question. "I don't know why I would," he says, speaking slowly. "I like my privacy." He's never seen Hannah's Facebook profile. "I haven't gone on it. I don't know how to get into it!" I ask him if he takes pictures when he attends parties, and he looks at me like I have three heads. "There are a lot of weirdos out there," he emphasizes. "There are a lot of strangers out there."

There is plenty of variation among this younger cohort, including a set of Luddite dissenters: "If I want to contact someone, I'll write them a letter!" grouses Katherine Gillespie, a student at Hunter College. (Although when I look her up online, I find that she too has a profile.) But these variations blur when you widen your view. One 2006 government study—framed, as such studies are, around the stranger-danger issue—showed that 61 percent of 13- to 17-year-olds have a profile online, half with photos. A recent Pew Internet Project study put it at 55 percent of 12- to 17-year-olds. These numbers are rising rapidly.

It's hard to pinpoint when the change began. Was it 1992, the first season of *The Real World*? (Or maybe the third season, when cast members began to play to the cameras? Or the seventh, at which point the seven strangers were so media savvy there was little difference between their being totally self-conscious and utterly unself-conscious?) Or you could peg the true beginning as that primal national drama of the Paris Hilton sex tape, those strange weeks in 2004 when what initially struck me as a genuine and indelible humiliation—the kind of thing that lost former Miss America Vanessa Williams her crown 20 years earlier—transformed, in a matter of days, from a shocker into no big deal, and then into just another piece of publicity, and then into a kind of power.

But maybe it's a cheap shot to talk about reality television and Paris Hilton. Because what we're discussing is something more radical if only because it is more ordinary: the fact that we are in the sticky center of a vast psychological experiment, one that's only just begun to show results. More young people are putting more personal information out in public than any older person ever would—and yet they seem mysteriously healthy and normal, save for an entirely different definition of privacy. From their perspective, it's the extreme caution of the earlier generation that's the narcissistic thing. Or, as Kitty put it to me, "Why not? What's the worst that's going to happen? Twenty years down the road, someone's gonna find your picture? Just make sure it's a great picture."

And, after all, there is another way to look at this shift. Younger people, one could point out, are the only ones for whom it seems to have sunk in that the idea of a truly private life is already an illusion. Every street in New York has a surveillance camera. Each time you swipe your debit card at Duane Reade or use your MetroCard, that transaction is tracked. Your employer owns your e-mails. The NSA owns your phone calls. Your life is being lived in public whether you choose to acknowledge it or not.

So it may be time to consider the possibility that young people who behave as if privacy doesn't exist are actually the sane people, not the insane ones. For someone like me, who grew up sealing my diary with a literal lock, this may be tough to accept. But under current circumstances, a defiant belief in holding things close to your chest might not be high-minded. It might be an artifact—quaint and naive, like a determined faith that virginity keeps ladies pure. Or at least that might be true for someone who has grown up "putting themselves out there" and found that the benefits of being transparent make the risks worth it.

Shirky describes this generational shift in terms of pidgin versus Creole. "Do you know that distinction? Pidgin is what gets spoken when people patch things together from different languages, so it serves well enough to communicate. But Creole is what the children speak, the children of pidgin speakers. They impose rules and structure, which makes the Creole language completely coherent and expressive, on par with any language. What we are witnessing is the Creolization of media."

That's a cool metaphor, I respond. "I actually don't think it's a metaphor," he says. "I think there may actually be real neurological changes involved."

I'm crouched awkwardly on the floor of Xiyin Tang's Columbia dorm room, peering up at her laptop as she shows
me her first blog entries, a 13-year-old Xiyin's musings on
Good Charlotte and the perfidy of her friends. A Warhol
Marilyn print gazes over our shoulders. "I always find myself more motivated to write things," Xiyin, now 19, explains, "when I know that somebody, somewhere, might be
reading it."

From the age of 8, Xiyin, who grew up in Maryland,
kept a private journal on her computer. But in fifth grade,
she decided to go public and created two online periodicals:
a fashion 'zine and a newsletter for "stories and novellas and
whatnot." In sixth grade, she began distributing her journal
to 200 readers. Even so, she still thought of this writing as
personal.

"When I first started out with my Livejournal, I was
very honest," she remembers. "I basically wrote as if there
was no one reading it. And if people wanted to read it, then
great." But as more people linked to her, she became correspondingly self-aware. By tenth grade, she was part of a
group of about 100 mostly older kids who knew one another
through "this web of MySpacing or Livejournal or music
shows." They called themselves "The Family" and centered
their attentions around a local band called Spoont. When a
Family member commented on Xiyin's entries, it was a
compliment; when someone "Friended" her, it was a bigger
compliment. "So I would try to write things that would not
put them off," she remembers. "Things that were not silly. I
tried to make my posts highly stylized and short, about
things I would imagine people would want to read or com-
ment on."

Since she's gone to college, she's kept in touch with friends through her journal. Her romances have a strong online component. But lately she's compelled by a new aspect of her public life, what she calls, with a certain hilarious spokeswoman-for-the-cause effect, the "party-photo phenomenon." Xiyin clicks to her Facebook profile, which features 88 photos. Some are snapshots. Some are modeling poses she took for a friend's portfolio. And then there are her MisShapes shots: images from a popular party in Tribeca, where photographers shoot attendees against a backdrop. In these photos, Xiyin wears '80s fashions—a thick belt and an asymmetrical top that give me my own high-school flashback—and strikes a world-weary pose. "To me, or to a lot of people, it's like, why go to a party if you're not going to get your picture taken?"

Among this gallery, one photo stands out: a window-view shot of Xiyin walking down below in the street, as if she'd been snapped by a spy camera. It's part of a series of "stalker photos" a friend has been taking, she informs me: He snaps surreptitious, paparazzi-like photos of his friends and then uploads them and "tags" the images with their names, so they'll come across them later. "Here's one where he caught his friend Hannah talking on the phone."

Xiyin knows there's a scare factor in having such a big online viewership—you could get stalked for real, or your employer could bust you for partying. But her actual experience has been that if someone is watching, it's probably a good thing. If you see a hot guy at a party, you can look up his photo and get in touch. When she worked at American Apparel, management posted encouraging remarks on employee MySpace pages. A friend was offered an internship by a magazine's editor-in-chief after he read her profile. All sorts of opportunities—romantic, professional, creative—seem to Xiyin to be directly linked to her willingness to reveal herself a little.

When I was in high school, you'd have to be a megalomaniac or the most popular kid around to think of yourself as having a fan base. But people 25 and under are just being realistic when they think of themselves that way, says media researcher Danah Boyd, who calls the phenomenon "invisible audiences." Since their early adolescence, they've learned to modulate their voice to address a set of listeners that may shrink or expand at any time: talking to one friend via instant message (who could cut and paste the transcript), addressing an e-mail distribution list (archived and accessible years later), arguing with someone on a posting board (anonymous, semi-anonymous, then linked to by a snarky blog). It's a form of communication that requires a person to be constantly aware that anything you say can and will be used against you, but somehow not to mind.

This is an entirely new set of negotiations for an adolescent. But it does also have strong psychological similarities to two particular demographics: celebrities and politicians, people who have always had to learn to parse each sentence they form, unsure whether it will be ignored or redound into sudden notoriety (Macaca!). In essence, every young person in America has become, in the literal sense, a public figure. And so they have adopted the skills that celebrities learn in order not to go crazy: enjoying the attention instead of fighting it—and doing their own publicity before somebody does it for them.

CHANGE 2: THEY HAVE ARCHIVED THEIR
ADOLESCENCE

I remember very little from junior high school and high school, and I've always believed that was probably a good thing. Caitlin Oppermann, 17, has spent her adolescence making sure this doesn't happen to her. At 12, she was blogging; at 14, she was snapping digital photos; at 15, she edited a documentary about her school marching band. But right now the high school senior is most excited about her first "serious project," caitlinoppermann.com. On it, she lists her e-mail and AIM accounts, complains about the school's Web censors, and links to photos and videos. There's nothing racy, but it's the type of information overload that tends to terrify parents. Oppermann's are supportive: "They know me and they know I'm not careless with the power I have on the internet."

As we talk, I peer into Oppermann's bedroom. I'm at a café in the West Village, and Oppermann is in Kansas City—just like those Ugg girls, who might, for all I know, be linked to her somehow. And as we talk via iChat, her face floats in the corner of my screen, blond and deadpan. By swiveling her webcam, she gives me a tour: her walls, each painted a different color of pink; storage lockers; a subway map from last summer, when she came to Manhattan for a Parsons design fellowship. On one wall, I recognize a peace banner I've seen in one of her videos.

I ask her about that Xanga, the blog she kept when she was 12. Did she delete it?

"It's still out there!" she says. "Xanga, a Blogger, a Facebook, my Flickr account, my Vimeo account. Basically, what I do is sign up for everything. I kind of weed out what I like." I ask if she has a MySpace page, and she laughs and gives me an amused, pixellated grimace. "Unfortunately I do! I was so against MySpace, but I wanted to look at people's pictures. I just really don't like MySpace. 'Cause I think it's just so . . . I don't know if *superficial* is the right word. But plastic. These profiles of people just parading themselves. I kind of have it in for them."

Oppermann prefers sites like Noah K Everyday, where a sad-eyed, 26-year-old Brooklyn man has posted a single photo of himself each day since he was 19, a low-tech piece of art that is oddly moving—capturing the way each day brings some small change. Her favorite site is Vimeo, a kind of hipster YouTube. (She's become friends with the site's creator, Jakob Lodwick, and when she visited New York, they went to the Williamsburg short-film festival.) The videos she's posted there are mostly charming slices of life: a "typical day at a school," hula-hooping in Washington Square Park, conversations set to music. Like Oppermann herself, they seem revelatory without being revealing, operating in a space midway between behavior and performance.

At 17, Oppermann is conversant with the conventional wisdom about the online world—that it's a sketchy bus station packed with pedophiles. (In fact, that's pretty much the standard response I've gotten when I've spoken about this piece with anyone over 39: "But what about the perverts?" For teenagers, who have grown up laughing at porn pop-ups and the occasional instant message from a skeezy stranger, this is about as logical as the question "How can you move to New York? You'll get mugged!") She argues that when it comes to online relationships, "you're getting what you're being." All last summer, as she bopped around downtown Manhattan, Oppermann met dozens of people she already knew, or who knew her, from online. All of which means that her memories of her time in New York are stored both in her memory, where they will decay, and on her site, where they will not, giving her (and me) an unsettlingly crystalline record of her 17th summer.

Oppermann is not the only one squirreling away an archive of her adolescence, accidentally or on purpose. "I have a logger program that can show me drafts of a paper I wrote three years ago," explains Melissa Mooneyham, a graduate of Hunter College. "And if someone says something in instant message, then later on, if you have an argument, you can say, 'No, wait: You said *this* on *this* day at *this* time.'"

As for that defunct Xanga, Oppermann read it not long ago. "It was interesting. I just look at my junior high self, kind of ignorant of what the future holds. And I thought, *You know, I don't think I gave myself enough credit: I'm really witty!*" She pauses and considers. "If I don't delete it, I'm still gonna be there. My generation is going to have all this history; we can document anything so easily. I'm a very sentimental person; I'm sure that has something to do with it."

CHANGE 3: THEIR SKIN IS THICKER THAN YOURS

The biggest issue of living in public, of course, is simply that when people see you, they judge you. It's no wonder Paris Hilton has become a peculiarly contemporary role model, blurring as she does the distinction between exposing oneself and being exposed, mortifying details spilling from her at regular intervals like hard candy from a piñata. She may not be likable, but she offers a perverse blueprint for surviving scandal: Just keep walking through those flames until you find a way to take them as a compliment.

This does not mean, as many an apocalyptic Op-Ed has suggested, that young people have no sense of shame. There's a difference between being able to absorb embarrassment and not feeling it. But we live in a time in which humiliation and fame are not such easily distinguished quantities. And this generation seems to have a high tolerance for what used to be personal information splashed in the public square.

Consider Casey Serin. On Iamfacingforeclosure.com, the 24-year-old émigré from Uzbekistan has blogged a truly disastrous financial saga: He purchased eight houses in eight months, looking to "fix 'n' flip," only to end up in massive debt. The details, which include scans of his financial documents, are raw enough that people have accused him of being a hoax, à la YouTube's Lonelygirl15. ("Foreclosure-Boy24," he jokes.) He's real, he insists. Serin simply decided that airing his bad investments could win him helpful feedback—someone might even buy his properties. "A lot of people wonder, 'Aren't you embarrassed?' Maybe it's naive, but I'm not going to run from responsibility." Flaming commenters don't bug him. And, ironically, the impetus for the site came when Serin was denied a loan after a lender discovered an earlier, friends-only site. Rather than delete it, he swung the doors open. "Once you put something online, you really cannot take it back," he points out. "You've got to be careful what you say—but once you say it, you've got to stand by it. And the only way to repair it is to continue to talk, to explain myself, to see it through. If I shut down, I'm at the mercy of what other people say."

Any new technology has its victims, of course: the people who get caught during that ugly interregnum when a technology is new but no one knows how to use it yet. Take "Susie," a girl whose real name I won't use because I don't want to make her any more Googleable. Back in 2000, Susie filmed some videos for her then-boyfriend: she stripped, masturbated, blew kisses at the webcam—surely just one of many to use her new computer this way. Then someone (it's not clear who, but probably her boyfriend's roommate) uploaded the videos. This was years before YouTube, when Kaazaa and Morpheus ruled. Susie's films became the earliest viral videos and turned her into an accidental online porn star, with her own Wikipedia entry.

When I reached her at work, she politely took my information down and called back from her cell. And she told me that she'd made a choice that she knew set her outside her own generation. "I never do MySpace or Facebook," she told me. "I'm deathly afraid to Google myself." Instead, she's become stoic, walling herself off from the exposure. "I've had to choose not to be upset about it because then I'd be upset all the time. They want a really strong reaction. I don't want to be that person."

She had another option, she knows: She could have embraced her notoriety. "I had everyone calling my mom: Dr. Phil, Jerry Springer, *Playboy*. I could have been like Paris Hilton, but that's not me. That thing is so unlike my personality; it's not the person I am. I guess I didn't think it was real." As these experiences become commonplace, she tells me, "it's not going to be such a big deal for people. Because now it's happened to a million people."

And it's true that in the years since Susie's tapes went public, the leaked sex tape has become a perverse, established social convention; it happens at every high school and to every B-list celebrity. At Hunter College last year, a student named Elvin Chaung allegedly used Facebook accounts to blackmail female students into sending him nude photos. In movies like *Road Trip,* "oops porn" has become a comic convention, and the online stuff regularly includes a moment when the participant turns to the camera and says, "You're not going to put this online, are you?"

But Susie is right: For better or worse, people's responses have already begun to change. Just two years after her tapes were leaked, another girl had a tape released on the internet. The poster was her ex, whom we'll call Jim Bastard. It was a parody of the MasterCard commercial: listing funds spent on the relationship and then his "priceless" revenge for getting dumped—a clip of the two having sex. (To the casual viewer, the source of the embarrassment is somewhat unclear: The girl is gorgeous, and the sex is not all that revealing, while the boy in question is wearing socks.) Then, after the credits, the money shot: her name, her e-mail addresses, and her AIM screen names.

Like Susie, the subject tried, unsuccessfully, to pull the video off-line; she filed suit and transferred out of school. For legal reasons, she wouldn't talk to me. But although she's only two years younger than Susie, she hasn't followed in her footsteps. She has a MySpace account. She has a Facebook account. She's planned parties online. And shortly after one such party last October, a new site appeared on MySpace: seemingly a little revenge of her own. The community is titled "The Society to Chemically Castrate Jim Bastard," and it features a picture of her tormentor with the large red letters *loser* written on his forehead—not the most high-minded solution, perhaps, but one alternative to retreating for good.

Like anyone who lives online, Xiyin Tang has been stung a few times by criticism, like the night she was reading BoredatButler.com, an anonymous Web site posted on by Columbia students, and saw that someone had called her "pathetic and a whore." She stared at her name for a while, she says. "At first, I got incredibly upset, thinking, *Well now, all these people can just go Facebook me and point and form judgments.*" Then she did what she knew she had to do: She brushed it off. "I thought, *Well, I guess you have to be sort of honored that someone takes the time to write about you, good or bad.*"

I tell Xiyin about Susie and her sex tape. She's sympathetic with Susie's emotional response, she says, but she's most shocked by her decision to log off entirely. "My philosophy about putting things online is that I don't have any secrets," says Xiyin. "And whatever you do, you should be able to do it so that you're not ashamed of it. And in that sense, I put myself out there online because I don't care—I'm proud of what I do, and I'm not ashamed of any aspect of that. And if someone forms a judgment about me, that's their opinion.

"If that girl's video got published, if she did it in the first place, she should be thick-skinned enough to just brush it off," Xiyin muses. "I understand that it's really humiliating and everything. But if something like that happened to me, I hope I'd just say, well, that was a terrible thing for a guy to do, to put it online. But I did it, and that's me. So I am a sexual person, and I shouldn't have to hide my sexuality. I did this for my boyfriend just like you probably do this for your boyfriend, just that yours is not published. But to me, it's all the same. It's either documented online for other people to see or it's not, but either way you're still doing it. So my philosophy is, why hide it?"

FUTURE SHOCK

For anyone over 30, this may be pretty hard to take. Perhaps you smell brimstone in the air, the sense of a devil's bargain: Is this what happens when we are all, eternally, onstage? It's not as if those '50s squares griping about Elvis were wrong, after all. As Clay Shirky points out, "All that stuff the elders said about rock and roll? They pretty much nailed it. Miscegenation, teenagers running wild, the end of marriage!"

Because the truth is, we're living in frontier country right now. We can take guesses at the future, but it's hard to gauge the effects of a drug while you're still taking it. What happens when a person who has archived her teens grows up? Will she regret her earlier decisions, or will she love the sturdy bridge she's built to her younger self—not to mention the access to the past lives of friends, enemies, romantic partners? On a more pragmatic level, what does this do when you apply for a job or meet the person you're going to marry? Will employers simply accept that everyone has a few videos of themselves trying to read the Bible while stoned? Will your kids watch those stoner Bible videos when they're 16? Is there a point in the aging process when a person will want to pull back that curtain—or will the My-Space crowd maintain these flexible, cheerfully thick-skinned personae all the way into the nursing home?

And when you talk to the true believers, it's hard not to be swayed. Jakob Lodwick seems like he shouldn't be that kind of idealist. He's Caitlin Oppermann's friend, the co-founder of Vimeo and a cocreator of the raunchy College Humor.com. Lodwick originated a popular feature in which college girls post topless photos; one of his first online memories was finding Susie's videos and thinking she seemed like the ideal girlfriend. But at 25, Lodwick has become rather sweetly enamored of the uses of video for things other than sex. His first viral breakthrough was a special effects clip in which he runs into the street and appears to lie down in front of a moving bus—a convincing enough stunt that MSNBC, with classic older-generation cluelessness, used it to illustrate a segment about kids doing dangerous things on the internet.

But that was just an ordinary film, he says: no different from a TV segment. What he's really compelled by these days is the potential for self-documentation to deepen the intimacy of daily life. Back in college, Lodwick experimented with a Web site on which he planned to post a profile of every person he knew. Suddenly he had fans, not just of his work but of him. "There was a clear return on investment when I put myself out there: I get attention in return. And it felt good." He began making "vidblogs," aiming his camera at himself and then turning it around to capture "what I'd see. I'd try to edit as little as possible so I could catch, say, a one-second glimpse of conversation. And that was what resonated with people. It was like they were having a dream that only I could have had, by watching this four or five minutes. Like they were remembering my memories. It didn't tell them what it was like to hang out with me. It showed them what it was like to be me."

This is Jakob's vision: a place where topless photos are no big deal—but also where everyone can be known, simply by making him- or herself a bit vulnerable. Still, even for someone like me who is struggling to embrace the online world, Lodwick's vision can seem so utopian it tilts into the impossible. "I think we're gradually moving away from the age of investing in something negative," he muses about the crueler side of online culture. "For me, a fundamental principle is that if you like something, you should show your love for it; if you don't like it, ignore it, don't waste your time." Before that great transition, some Susies will get crushed in the gears of change. But soon, he predicts, online worlds will become more like real life: Reputation will be the rule of law. People will be ashamed if they act badly, because they'll be doing so in front of all 3,000 of their friends. "If it works in real life, why wouldn't it work online?"

If this seems too good to be true, it's comforting to remember that technology always has aftershocks. Surely, when telephones took off, there was a mourning period for that lost, glorious golden age of eye contact.

Right now the big question for anyone of my generation seems to be, endlessly, "Why would anyone do that?" This is not a meaningful question for a 16-year-old. The benefits are obvious: The public life is fun. It's creative. It's where their friends are. It's theater, but it's also community: In this linked, logged world, you have a place to think out loud and be listened to, to meet strangers and go deeper with friends.

And, yes, there are all sorts of crappy side effects: the passive-aggressive drama ("you know who you are!"), the shaming outbursts, the chill a person can feel in cyberspace on a particularly bad day. There are lousy side effects of most social changes (see feminism, democracy, the creation of the interstate highway system). But the real question is, as with any revolution, which side are you on?

Charles Graeber

The Pedal-to-the-Metal, Totally Illegal, Cross-Country Sprint for Glory

One man's quest to set a new record for the coast-to-coast Cannonball Run

And so the clock starts and the taillights flare, and they're off again, strapped down, fueled up, and bound on an outlaw enterprise with 2,795 miles of interstate and some 31,000 highway cops between them and the all-time speed record for crossing the American continent on four wheels.

The gear is all bought and loaded. Twenty packs of Nat Sherman Classic Light cigarettes, check. Breath mints, check. Glucose and guarana, Visine and riboflavin, Gatorade and Red Bull, mail-order porta-pissoir bags of quick-hardening gel, check.

Randolph highway patrol sunglasses, 20-gallon reserve fuel tank, Tasco 8 × 40 binoculars fitted with a Kenyon KS-2 gyro stabilizer, military spec Steiner 7 × 50 binoculars, Hummer H1-style bumper-mounted L-3 Raytheon Night-Driver thermal camera and LCD dashboard screens, front-and rear-mounted sensors for a Valentine One radar/laser detector, flush bumper-mount Blinder M40 laser jammers, redundant Garmin StreetPilot 2650 GPS units, preprogrammed Uniden police radio scanners, ceiling-mount

Uniden CB radio with high-gain whip antenna. Check. Check. Check.

At the moment, the driver and copilot of this E39 BMW M5 are illegal in intent only as they obediently cow along the tip of Manhattan, funnel into the Holland Tunnel, and spill out into New Jersey along a six-lane mash-and-merge. The speedometer reads a cool 60 miles per hour; the clock reads 9:12 p.m.

"Unacceptable," Alex Roy says. The 35-year-old driver is addressing both the numbers and himself. Then, after 20 sickening minutes in construction traffic, Roy says it to the darkened highway, pushing up over 110 mph while his copilot squints along the scabbed blacktop for the deer that might end their lives and the policemen who might kill their trip.

The quest itself—to cross from New York to Los Angeles with unthinkable brevity—is a drive, yes, in the same way that the moon shot was a flight. This is an engineered operation that has been financed, scarioed, calculated, technologically outfitted, and (via digital video and triangulated time-stamped texting and GPS verification and support teams on both coasts) will be monitored and recorded (for proof, posterity, and a documentary film).

For nearly two years, Roy—a pale, shaved-headed, independently wealthy ectomorphic veteran of the Gumball 3000 road rally—has obsessed sleeplessly over every detail and thrown money at every possible electronic connivance. His mission is intended as a triumph of the mind over the base adrenal impulses of common speeders. His route is nothing like the careless line a spring breaker might plot across a Rand McNally—it's a painstakingly GPS-mapped and Google Earth–practiced manifest desti-document, waypointed mile by mile for detours, construction, and speed traps.

White lines scroll through the windshield and mile markers tick past the tires as Roy flips a series of toggles on the center console, killing the brake lights (to prevent telltale flashes if he needs to slow for sudden radar), then flips a few more to illuminate the cockpit with night-vision-friendly red LEDs. The cockpit glows like a submarine at battle stations. Now Roy punches up the digital codes corresponding to the New Jersey State Police on the police scanner. The car fills with the coded squawk of emergency dispatchers, speeding motorcycles, and domestic quarrels.

"OK, scanner is live," Roy says. He hits another switch under the dash, and a light goes green on his steering wheel display. It means that the vehicle is now traveling in a sort of force field of infrared light, a bubble that deforms the bandwidth of incoming police laser spotters. "Jammers are active," Roy says. "Now let's have the radar."

Roy's current copilot, an English racer named Henry Fyshe, reaches under the seat and pulls out the Valentine One. He plugs it into the bank of fused circuits snaking from the car's power supply and flips the switch, and now another instrument joins the cacophony. The Valentine picks up incoming radar: mostly the X and K bandwidths. The bleeps of X-band are usually just junk picked up from motion detectors and burglar alarms and the shipping docks of Port Elizabeth to the south. But the occasional croaking *blaaat!* means K-band—and almost certainly a police trigger gun hitting home.

The combination of *bleep! bleep! blaat! bleep!* is chaos pinpricked with information. Listening, sorting, interpreting—it's all exhausting. Then Roy reaches overhead and flips on the CB, adding an overlay of truck driver patois: twangy talk of big-boobie women and fishing and traffic on the I-78.

"Fascinating," Fyshe says. Compared with the thick

southern drawl coming from the speaker, his polished Oxbridge English sounds as refined as drawing room French.

"OK, CB is active," Roy says above the noise. "Now check the thermals, please, Mr. Fyshe. We need to start banking time."

There's something very Captain Jean-Luc Picard about Roy. Maybe it's the top-gun lingo and ramrod driving posture. Maybe it's his bald, ovoid skull or his habit of wearing faux military uniforms during races. Or maybe it's because Roy is actually in command of his very own road-bound USS *Enterprise*. Captain Roy is determined to boldly go faster than any man has gone before.

Roy is attempting to break a legendary cross-country driving record known to most people as the Cannonball Run. The time: 32 hours, 7 minutes, set in 1983 by David Diem and Doug Turner. Captain Roy's quest is definitely illegal and quite possibly impossible. He is one of the few drivers wealthy and geeky and foolish enough to try it anyway. So far he's tried and failed twice, but he's still convinced that his careful calculations will allow him to beat the record.

At the core of his plan are his beloved spreadsheets. Roy, with help from a car-crazy former New Jersey transportation department employee named J. F. Musial, has spent months loading Excel documents with the coordinates of all-night gas stations and open stretches of highway and weather projections—hundreds of data points arranged on an x-y axis, so that any deviation can be recalculated on the fly.

The resulting document is as thick as a stock prospectus—and just as unreadable, particularly if you're driving in the dark at 50 mph over the speed limit. But the security blanket of overclocked data calms Roy. It's his hedge against all the uncertainty and risk—of vehicular homicide, of jail time, of failure. Racing across the country is a foolish and dangerous and ill-advised dream, and Roy knows it.

But after more than a year of bitter experience, Roy has discovered that even an *Enterprise*'s worth of Excel spreadsheets can't control the weather or the traffic or the deer or the possibility of mechanical failure. Or the police—especially the police.

So far his failed attempts to beat the record have cost Roy a lot of time and money, at least one girlfriend, and even his original, trusted copilot. Instead of glory, Roy's cross-country trips have brought him a mechanical breakdown, a police investigation, multiple radio alerts, and one arrest. And with each setback, Roy risks blowing the secrecy of his quest and putting the brakes on forever. He is quickly running out of chances to drive his dream. If he's going to beat 32:07, he'd better do it soon.

He's hoping Fyshe is the right partner. Like Roy, Fyshe is wealthy and single and an excellent driver. Unfortunately, he's also far more experienced steering his immaculate 1954 OSCA MT4 Maserati through Italy's Mille Miglia endurance race than dodging minivans along Jersey's I-78. Roy is stuck in the middle of a criminal automotive enterprise with a copilot who can't spot an American cop.

"OK," Roy says. "Now, see that?"

Fyshe frowns and peers through the windshield at a dark American town car.

"That's never a cop," Roy says. "Just a taxi."

Fyshe nods, intrigued. "I see," he says.

"Now, see that?" Roy points out a yellow cab, just visible in the distance. "The taxi? That's the type of car."

"It's a taxi?" Fyshe asks.

"Yes, it's a taxi," Roy says. "But in a dark color, that can be an unmarked cop."

"How can you tell the difference?" Fyshe asks.

"You just have to," Roy says.

"I see," Fyshe says. But he doesn't, not really.

Roy gives it the gas, easing up toward 90 mph, passing two trucks, flashing by a Corvette in the slow lane, and pushing up a hill at 93. "Ramp check?"

Fyshe glances reflexively to the right and studies the cars pouring down the entrance ramp, looking for lights on top. "Clear."

"Now, see that overpass ahead?" At 100 mph now, it's approaching fast. "Check the thermals."

Fyshe checks the dash, where the bumper-mounted night-vision camera feeds a thermal image to a seven-inch dashboard display. The traffic ahead glows in the darkness like the Predator.

"If a cop is idling around one of those columns, he'll have his engine on and show up as heat," Roy says. "Unless there's a concrete barrier that shields him. Check the sheet."

Roy feels into the side pocket and hands Fyshe a series of color-coded sheets. "Barriers—yes, except where marked by DOT signs," Fyshe reads.

"It also says the limit is 65 mph here," Fyshe says. "What are we now?"

"Ninety-eight."

"Jolly good," Fyshe says, delighted. "But what if there's a policeman on top of one of those bridges?"

"It's an overpass," Roy says. "And there won't be."

"Cameras?"

"Nope," Roy says. "The plate covers reflect flash anyway."

"In Europe, there are cameras everywhere," Fyshe says thoughtfully. "The police see everything." He watches the white lines blur into a continuous streak, lost in the Wild West of central Jersey.

The highway crosses the state in an undulating sine wave. At each new rise, Fyshe scans the thermals ahead and glances behind to the ramp before Roy punches the clear valley at 100 mph, bringing the trip average up to 82.3 mph. This is the Jersey nobody ever thinks of—empty, three lanes, no traffic or stores or malls—so when K-band suddenly croaks on the scanner, Roy knows it's no false alarm.

"Where are you?" he mutters. A red arrow glows on his steering column, meaning radar from ahead.

"If he's behind us and not in sight, hit the gas," he tells Fyshe. "If he's ahead, ease off until you establish position."

Roy crests the hill, eases off the gas, and takes the right lane. He's just a law-abiding citizen now. Standard police protocol is for a cruiser to lie at the side of the road just over the crest of a hill, exactly when drivers have their foot on the gas and no view ahead. By taking the right lane, a speeder approaches a radar gun with the sharpest parallax angle— the least accurate for getting a clean read.

"I don't see him," Roy says. "I'll take this hill easy and—"

Blaaat! goes the scanner. *Blatt! Blaat!* Sure enough, the downhill is lit by the strobing rack lights of a New Jersey state trooper, ringing up some poor schmuck in a minivan.

"Now, that's a cop," Roy says. He hits a button on the GPS unit's touch screen, adding yet more data—the location of this speed trap—before confidently stepping back on the gas.

Going cross-country fast is not rocket science, but in Roy's world it does require a lot of basic math. To beat the record, Roy has calculated that he needs to maintain an average of almost exactly 90 mph from Manhattan to the Santa Monica Pier. For occasional spurts, 90 is not uncommon on the highway. But for a day and a half of barreling across the United States, 90 miles per hour is essentially insane.

As a Cannonballer makes his way across the continent, the accumulation of his time and speed forms a rising and falling curve called a running average. For every second spent below his 90 mph target, Roy will need to compensate by investing a second going faster than that average. Which is why Roy doesn't want to stop. Every second spent at 0 mph is a second he can never recover—even with his BMW's factory-set 155 mph limiter replaced with a Power-chip ECU engine chip. Unfortunately for Roy, no matter how carefully he keeps to his fuel-efficiency regimen or how large his spare fuel tank, he will need to pull over and gas up at least five times.

Then there's the weather—projected to be nasty from Indianapolis to St. Louis, at least—and the reality that every 12 hours the rest of America will pack into their PT Cruisers and steer directly onto Roy's racetrack. The only way Roy and his copilot can even hope to average 90 mph is to plan (Roy has, fanatically), pray (a friend petitioned a Taoist spiritual master for them), and, wherever possible, stomp the throttle (they are).

The trip has just begun, but Roy is already in trouble. There's a closed gas station he hadn't foreseen and that surprise construction in New Jersey—not to mention a green copilot unfamiliar with American cop customs. Each small deviation from the plan ripples through the rest of the spreadsheet. His calculations are already starting to crumble, and Roy's 72 mph cumulative average is pathetically low. He needs to put time in the bank.

He grabs the CB mic. "Breaker breaker, I need a bear check, over," he calls.

"Yeah, you're clear on the 78 all the way to the Buckeye," comes the voice, and Roy punches it, hitting 130 along a black stretch of road as the topography becomes hillier, the trees leafier. He's brought the average up to 78.4 by 1:00 a.m.

and 80 by 2:00 a.m., when the BMW barrels through a tunnel and flicks across trestle bridges into Ohio—the most famously perilous state for speeders.

"Switch the scanner frequencies immediately!" Roy says, and sure enough the CB starts crackling with word of Smokies rolling westbound, then two more in the hammer lane, one with a package, another in a plain brown wrapper, now trailing just a half a mile marker behind Roy. Only unreasonable speed can put distance between them, so Roy takes the CB mic. "Breaker breaker, can I get a bear check?" he calls again.

"Bear check? That something they teach you in trucker school?" comes the answer.

It's nearly 4:00 a.m. Roy gasses through Columbus, then Springfield. The billboards snap past the windows like the pages in a flip book. By 4:30, the speedometer shows a steady 102 mph, but the overall average is only 82. It's far too slow to break the record. At this point, it's impossible to bring it back up.

"I'm calling it," Roy sighs. "That's it." And so, at 4:20 in the morning, some 70 miles shy of the Indianapolis Motor Speedway, Roy puts his turn signal on like some average commuter and once again stops, 2,160 miles short of his dream.

Alex Roy's Cannonball dreams started with a movie, but it didn't star Burt Reynolds. At the time, the 27-year-old Roy was living in New York after his father had called him back from Paris, where Roy had been working part-time at a bar and trying to write the Great American Novel—set, arbitrarily, in Japan. His father was in the hospital, sick with throat cancer, and Roy had traded in his life as an artiste to manage the family business, a rental agency called Europe By Car. The young heir was at sea, fresh from an unsuccessful attempt to forge his own identity and sitting in a trendy Soho bar-cum-theater called Void. And then the lights went down, and Roy saw the future.

The film was *C'était un Rendez-vous*. Made in 1976, it's a dashing precursor to every *Jackass*-inspired digicam stunt ever posted on YouTube—nine heart-pounding minutes choreographed to a screaming drivetrain. Through a bumper-mounted camera, the viewer becomes the car— traveling more than 80 mph as the anonymous driver revs into the enormous traffic circle around Paris's Arc de Triomphe, steers hammer-down from the Champs Élysées to Sacré-Coeur in Montmartre (through 16 red lights, wrongway one-ways, stunned pedestrians, garbage trucks, and median strips) to meet up with a beautiful blond waiting patiently in the park at the Montmartre church.

Roy left Void in a state of dazed revelation. From a public safety perspective, he says, he knew *Rendez-vous* was just short of "a snuff film on wheels." But it was also the single coolest thing he'd ever seen.

The film's unmasked director and driver, Claude Lelouch, eventually achieved immortal fame and respect on the internet, fueled in part by old reports that Lelouch had been arrested after the film's first screening. Standing in a bar on a summer's night, a life as a feckless novelist behind him, another of trying to fill his father's wingtips ahead of him, Roy began to wonder: Could he make his own *Rendez-vous*—in New York? Could he be the great driver, mastering the city and meeting the blond?

He approached the question with a formula he'd repeat throughout his driving career. First he obsessed, talking ad nauseam about Lelouch's film to anyone who would listen. Then he drove his route repeatedly in his Audi S4, meticulously recording potholes and potential speed traps, then studying the lists on a color-coded cheat sheet. He planned to recruit close friends from his Manhattan private high school days to impersonate orange-vested traffic police to block traffic on race day.

The original idea was to make a full lap of Manhattan (skipping the most northerly and heavily policed sections of the city) in 25 minutes. This meant running dozens of red lights at absurd speeds and left little time to react to sudden contingencies like pedestrians. The stunt was dangerous and illegal, its success dependent on secrecy. But Roy has no talent for keeping secrets, particularly about his daring. (He was, in fact, using most of the recon runs to impress women.) By the end of the year, dozens of people knew about Roy's plan to *Rendez-vous* Manhattan.

But while outlaw street racing may sound romantic, the reality of a 29-year-old with no experience skidding through the most populous urban center in America is terrifying, not to mention feloniously stupid. Even Roy's girlfriend refused to play her part of meeting him at the finish line. The idea of actually having to follow through with his big plans started keeping Roy up at night; but the humiliating prospect of backing down was just as bad.

In the end, Roy never attempted the 25-minute Manhattan *Rendez-vous*. But he claims to have raced a 27-minute "practice run." He proudly estimates that he hit top speeds of 144 mph while committing 151 moving violations— enough to have his New York driver's license suspended 78 times over. And afterward, Roy says, "I never felt better." He had missed his goal but found his identity. Roy wanted to be known as an outlaw driver.

The fastest way to his new goal was to enter a road rally inspired by yet another movie—the 1976 cult classic *The Gumball Rally*. The film depicted a madcap outlaw road race; its real-life version is a 3,000-mile celebrity-and-socialite-studded international road rampage first organized in Europe in 1999. There are no qualifying events, and no experience is required. Entrants need both flash (tricked-out Bentleys, Porsches, and Lamborghinis encouraged) and cash (£28,000—about $56,425—for the 2007 rally), as well as the ability to keep a straight face while agreeing to a code of conduct that explicitly prohibits breaking any laws—including the speed limit. But while most Gumballers are rich young men paying for 3,000 miles of silicon-bimboed pit stops and Vegas-weekend-style bad-boy hoo-ha, Roy was one of the few actually racing to win.

He impressed the 2003 Gumball entry committee by topping the already well-represented freak factor: He wore a pastiche of authentic international police outfits and drove a rare E39 BMW M5 he claimed was used by the elite German "Autobaun Interceptor Unit," complete with police sirens and stickers. Roy's "Polizei 144" shtick added yet another layer of slapstick to the Gumball's air of a movie-come-to-life. Roy established a reputation as a fun-loving clown who also happened to be a fast, safe driver. He was an instant hit with race fans. His Web site attracted a small but faithful following that bought $500 Polizei 144 racing jackets and downloaded clips from his "Spirit of the Gumball" trophy win in the 2003 run, held in the United States.

Most of the comments on his site were typical rock-on fan blurts, but one was a challenge to "check out the *real deal*." Roy followed a Web link and, stunned, met his newest dream.

Once again, it was a movie—this time a trailer for a documentary in progress titled *32 Hours 7 Minutes,* covering the transcontinental racing record set by Diem and Turner. Here was an automotive stunt that had remained unequaled for almost 22 years. Anyone who topped it would be guaranteed fame and street cred; for Roy, this was *Rendez-vous* déjà vu. He immediately called the filmmaker, a diminutive speed fanatic named Cory Welles. Roy had the funding—and the perfect ending for her movie.

Most people remember *The Cannonball Run* as a campy '80s road comedy featuring, among others, Roger Moore, Dom DeLuise, and Farrah Fawcett. But to gearheads, the Cannonball Run is the original outlaw cross-country road race, organized by legendary *Car and Driver* writer Brock Yates. Entrants drove everything from cheap beaters to high-priced tweakers, but all had an appetite for white lines, black tar, and speed.

Officially known as the Cannonball Baker Sea-to-Shining-Sea Memorial Trophy Dash (and later as the U.S. Express race), the race set the standard for outlaw driving. This was uniquely American car culture—free and fun and fast. And nobody was faster than Diem and Turner, who hammered their 308 Ferrari from a garage on Manhattan's Upper East Side to Newport Beach, California, in an unthinkable 32 hours and 7 minutes.

According to Yates and his fellow Cannonballers, trying to beat that record today is pointless. Their argument goes something like this: Cannonball records were set back when the free-wheelin' '70s hooked up with the greed-is-good '80s for fat lines of cocaine and unprotected sex. But these, brother, are Patriot Act days—executive-privilege endtimes in which no rogue deed goes untracked, no E-ZPass unlogged, no roaming cell phone unmonitored by perihelion satellite. Big Brother is definitely watching. Big Speed, the old Cannonballers say, is a quaint, 20th-century idea, like pay phones or print magazines.

But nobody had telexed Roy or his new filmmaker pal, Welles, the memo on this one. Once again, Roy put his formula in motion. First, he planned for weeks. Then, with his high school friend Jon Goodrich as copilot and cameraman James Petersmeyer tucked in the backseat, Roy left Manhattan's Classic Car Club on December 16, 2005, and drove west, fast. They arrived at the Santa Monica Pier in California bleary-eyed, exhausted, and frightened—and 2 hours and 39 minutes shy of the record.

Roy and Goodrich flew back to New York to revamp their calculations and tried again on April 1, 2006. They were zeroing in on the 32:07 space shot—until the car broke down in Oklahoma. Roy was devastated. He immediately began planning another run.

But this time, Roy returned to his calculations by himself. Two hairy cross-country runs had been more than enough for Goodrich, and he simply wasn't willing to continue risking life, limb, and liberty for another man's dream. By now, though, replacing his copilot was the least of Roy's Cannonball problems. Despite the nondisclosure agreements, word was getting around. Back in September 2005, Roy's bearded and bullying Gumball 3000 frenemy, Richard Rawlings, had bet him $25,000 on a cross-country race—and another $25,000 that Rawlings would do it in less than 25 hours.

Roy refused the challenge, but it clearly meant time was running out. Sooner or later, somebody was going to try to break that record. If they succeeded, went on Leno, stole the glory—that would be bad for Roy. But if they got caught trying, that was even worse. Roy was sure that the police would then crack down, and the window of opportunity for his cross-country sneak would slam shut forever.

In fact, that window was closing already. After so many high-speed cross-country runs, Roy wasn't famous—but his antics were. He was already well remembered in Arizona, where he'd been arrested for speeding during a 2004 rally called the Bullrun wearing jackboots, German police togs, and a regulation leather police belt with handcuffs. (The concerned police psychiatrist asked Roy, "Do you know what year this is?") Ohio presented another problem. While running nearly 120 mph in a 55 zone on the return trip from the aborted Cannonball run with the English copilot, he'd been hit with radar by a westbound state trooper, leading to a tense, 20-minute Smokey-Bandit chase deep into farm country. Roy managed to escape, but the Ohio state patrol would be unlikely to forget the blue BMW loaded with weird antennas.

Roy faced similar problems in Pennsylvania and Oklahoma. On the April 2006 trip, Pennsylvania police dispatch reported a BMW without taillights speeding down the interstate. Then, waiting in the airport after the Oklahoma breakdown, Roy made the mistake of running his mouth off on a cell phone. The traveler in line behind him couldn't help noticing the strange bald man and overhearing words like *night vision, escape, cops,* and *spotter plane.* He called in a potential homeland security threat.

Roy eventually made it home, but Oklahoma authorities tracked his car to the local BMW dealership. The cops impounded the vehicle—still loaded with GPS units documenting his street racing—for three days while they investigated Roy.

"Needless to say, my attorney wasn't pleased," Roy says. "Actually, I think *stupid* was the word he used."

By fall 2006, the run-ins had reached critical mass. Before long, Roy feared, state authorities would connect the dots and shut him down for good. Within a month, winter snow might kill his time, and spring might be too late. If Roy was going to break the record, it was now or never. But first, he needed a new copilot.

It's a typically rainy September evening, only nine days before his next scheduled departure, and Roy is bug-eyed, chain-smoking, and pacing the length of his 2,571-square-foot bachelor pad in Manhattan's Cooper Square while his race team waits on his L-shaped couch, drinking his liquor and watching *Battlestar Galactica* on a massive projector screen. Each surround-sound kinetic energy weapon rattles ice in the drinks.

Roy checks his watch and then his desk, where three GPS units and four computer screens each display the time. Standing with his hands on his hips in front of the rotating world-map screen saver, he looks less like Captain Picard and more like a chain-smoking Lex Luthor.

"It's not like him to be late," he says. "What if he's incapacitated or dead?"

In choosing a new copilot, Roy considered lots of drivers (including me), before finally settling on a straitlaced 32-year-old finance-sector type named Dave Maher. From the first meeting, it was obvious that Maher and Roy would make a particularly odd couple. Roy is a fast-talking geek, as dead-eyed serious about the patches he Velcros onto his race uniforms as a *Star Trek* reenacter is about having the right blades on his Klingon battle d'k'tagh. Maher is quiet and has never watched *Battlestar Galactica*. He likes sports involving inflatable balls and has a penchant for red wine and amateur track club events for his 1996 Porsche 911 Turbo.

But both of them wanted to go fast, and something that Maher mentioned when they talked about the cross-country attempt struck a chord deep within Roy: a need to have something "that money couldn't buy." Maher had the job, and the odd couple became a team.

Roy wears his phone on his belt like Batman or a paper-products salesman, and now it begins to vibrate. He snaps it to his ear. "We're all here waiting," he says to the doorman. "Yes, send him up."

Maher arrives in a suit and tie, a bottle of excellent wine in hand, ready for a civilized party. Instead, Roy hands him his latest timetable. It is the product of 150 hours of work, a whopper version of all previous calculations. Roy has titled it "31:39 Driveplan .9d (Merciless Assault Reprisal −11)."

He hands the stack to Maher, who flips through the pages. The copilot looks like a kid on the first day of summer facing a pile of required reading.

"Ultimately, this drive is a math calculation," Roy says. Maher looks blank. Roy points to a series of cells in the spreadsheet. Maher scans it, then turns the page, searching. "See," Roy says, "that's the average we're looking to hit: 90."

"I know this average," Maher says quietly. He flips through more pages. "I'm looking for the extended stretches of big speed, the long stretches where we can really hit it and make time."

Roy straightens. "Well, those don't really exist," he says. "You'll see. It's very rare to run over 100 for even a minute or two . . ."

"Oh yeah?" Maher says smiling. "Well, I'm about to change that."

And so, on the Friday before Columbus Day weekend, the clock is punched and the taillights flare, and Roy once again rolls through the Holland Tunnel and across New Jersey. They cross the empty tarmac of Pennsylvania and into Ohio, gas up maniacally, and are back on the highway with Maher now doing 120 through the most famously cop-heavy state in the union. By Akron they've been driving all night, and the trip is just beginning. More Red Bulls are popped, vitamins taken, cigarettes lit, and then comes the sun, shockingly bright. Roy finds the Visine, then trains his attention on the shaking landscape. This is a criminal game of I Spy, using binoculars designed for battle—Steiners with independently autofocusing lenses—but at Maher's speed they just beat uselessly against Roy's eye sockets.

"You know, I just have a very hard time spotting like this," Roy says.

"We have to bank time," Maher says.

"It's averaging 91.3 mph," Roy says. "The projections say we're good."

"Your projections are conservative," Maher says. His eyes never leave the road. He looks strangely relaxed doing 130 mph. The radar is exploding with undercover police, and yet he's doubling the speed limit for the sort of sustained periods that Roy knows are potentially fatal to this quest.

"We need to go as fast as possible, every chance we get," Maher says, glancing at Roy. "Otherwise, we are definitely not going to make it."

"OK," Roy says. But he doesn't mean it. Maher's stomach for risk isn't found anywhere on Roy's spreadsheets, and this is way outside his comfort zone. "But I'm telling you, Dave, you get caught and—"

Now the radio explodes with a fresh voice. "Cowbell Ground, Cowbell Ground, this is Cowbell Air, over."

"Yes!" Roy says. He grabs the mic. "Cowbell Air, this is Cowbell Ground, go ahead."

"We have a visual," the voice from above says. This is Roy's secret weapon, a small Beechcraft twin-engine spotter plane piloted by Paul Weismann, a high school friend, along with another pilot named Keith Baskett. They're scouting for cops, traffic, and construction during the vulnerable daylight drive across the Midwest.

"How are we looking, over?" Roy asks.

"You're looking very fast and very nice," comes the voice from above. "All clear, boys, put the hammer down."

Maher pushes the car, passing even the gutsiest speeders

at nearly double speed. The white line is a ticking blur, the overpasses are distant, then here, then gone, and Texas is just a flat fuzz in the rearview. Near Oklahoma City, they stop for the Chinese fire drill of piss, pump, and go, and now Roy takes the wheel again, gunning to fly. The GPS says that even with gas stops, they've crossed half the country at 93.6 mph.

The highway ahead is fairly open, but the left lane is not, and this time, inspired by Maher's driving or the average or both, Roy does what he needs to do to keep the pace—passing one car on the right, pushing inches from the bumper of a 16-wheeler, then cutting left again to take the lane. And as if on cue, a female voice cuts in on the police scanner. "Report of a blue BMW speeding, weaving in and out of traffic, and driving recklessly. Be advised, unable to get tags . . ."

"That's us!" Maher says.

"Shit!" Roy says.

He cuts the brake lights on the panel and slows to double digits.

"What do we do?"

"Well, we're stuck in traffic."

"Where do we hide?" Roy asks. The land is flat to the horizon.

"We don't hide anywhere," Maher says.

Blaaat! Now the cockpit fills with the awful croak of K-band from a dead-on police trigger radar. "God damn it, where is that guy?" Roy mutters, then suddenly sees him— an SUV highway patrol car headed eastbound, and no median between them.

"Oh my God, he's braking!" Roy shouts. "He's crossing! We have to get to the next exit and hide."

"I don't know if we're going to have a lot of room to hide out here," Maher says.

Roy glances back and forth, mirror to road and back

again. Already, he's soaked through his shirt, his bald head raining sweat onto his sunglasses. The exit is coming fast. "Should we get off?" he asks. "Should we get off right now?"

The scanner again, a male voice: "Blue BMW on up ahead of me."

Then another voice—a second car: "Dark-blue BMW, tinted windows—looks like it has some antennas on it."

"I'm going," Roy says. He pulls up the exit ramp, taking the rise, rolling the stop sign like a normal driver, nothing in his mirror yet, then moves quickly to the right.

But this time, there's no getting away. It's farmland, flat forever—*North by Northwest,* a house in the distance, animals. Roy pulls to the side. He hops out of the car. He unzips his fly.

"I'll tell him we had to piss," he yells.

The male voice on the scanner again. "They're ahead of me," it says.

Roy looks. Nothing. "Hey!" he says. "He thinks we're still going!"

Roy zips up and turns, and now he sees it: a black-and-white coming up the ramp behind him. "Oh no," he says. The car pauses at the top of the ramp, then turns toward him. "Here he comes . . ."

Sitting in the passenger seat, Maher now looks around at the piles of GPS units, the maps and plans and scanners, the squawking boxes. He's sitting in an electronic crime scene. "Maybe I should turn something off?" he asks.

"Turn it off, turn it all off!" Roy shouts. He reaches into the center console to kill the main power just as the police car approaches. "What the . . . ?"

It's a black-and-white, all right: one of those ad-wrapped VW bugs with a giant GEEK SQUAD sticker where the sheriff's star might be. Suddenly, the sweat on Roy's head is cool and soothing.

"Maher," Roy says, "how come you can drive like that for seven hours and no one calls, and I do it for three minutes and then someone calls?"

"Because I'm Irish," Maher says.

They're off the highway for a total of two minutes. Even with the time lost to a dead stop, their overall average on the GPS stands at 95.7 mph—well above record pace. But there are storm clouds on the horizon, which become hard rain by New Mexico. The traffic clots, and the smeared windshield glows red with truck lights. With the darkness, the rain becomes blinding, blunting the vision of the thermal cameras. They enter Arizona in traffic, with a soul-killing 22 mph on the GPS and a forest of lightning on the horizon.

Maher pounds the wheel in disbelief. "No!" he shouts. "I've been driving so hard. . . . No!" He cuts into the breakdown lane to make a desperate run for it. Even an unsafe pass isn't possible. "No!" he repeats.

Mile after mile, their hard-won average withers, and the adrenaline dies with it. The rain is impossible. Maher is exhausted. "Maybe I'm seeing stars," Maher says.

"No, you're seeing the real thing," Roy says. The weather is clearing.

By Arizona, the pavement is dry. Maher gives it his last surge of energy, climbing to 122 mph, 142, 160, before the gas light demands they stop for fuel. It's 12:03 a.m. local time. They've been on the road for 29 hours and 27 minutes. The effort of this last sprint has pushed Maher to the breaking point. He staggers from the car on failing legs. The Casio counts the seconds as Roy plugs in the nozzle and stands, tweaked and muttering in front of the mini-mart like a meth kid getting a Big Gulp.

"You're done," Roy says. He falls into the driver's side and guns it back onto the highway for the final 131-mile stretch from Barstow to the Santa Monica Pier.

"I'm not sure that we're going to make it now," Maher says. His fingers fumble with Roy's projection chart, suddenly interested, but it's an unintelligible jumble of numbers. "You'll have to be above 100 the whole time, or we've driven a day and a half for nothing."

"I've got it," Roy says. He stares ahead like a machine. "Just watch the road."

After 7,700 miles and three attempts to cross the country at warp speed, Captain Roy has experienced something like a Maher mindmeld. As in any marathon, exhaustion and fear make quitting seem smart. You can say you tried, blame the weather, and find a hotel. But breaking a record—any record—takes something more, something personal. Right now, it will take everything. There's no room left for strategy. Roy simply has to hit it hard.

The radar is crazy with *bleep!* and *blatt!* The spreadsheets litter the cockpit like dirty floor mats, but Roy hits it anyway. He doesn't need charts anymore. He is the chart and Excel and Google Earth and Garmin MapSource and something more, too, a guy with something to prove.

He passes a minivan in the car pool lane at 102 mph and merges onto California's I-10 headed into Los Angeles with blocks of lit towers to the right and oncoming halogens kaleidoscoping his bleary corneas. But Roy sees only the road ahead and the best path through it, the racing line that shaves fat off the hips of the curves as he apexes them at 100 mph, now 117 past Crenshaw Boulevard, La Brea Avenue at 115. The curve and acceleration is a physical sensation in the gut, and now the city is 10 miles out, now 8, and the turbos spool up and kick and Maher says, "Cop! No—taxi!" while Roy hits 117 past Cloverfield Boulevard, peels off on the exit to a light gone green, the next one green—one, two, three more—through the gate of the Santa Monica Pier, where wooden planks rattle beneath the car.

It feels weird to slow, crazy to stop, but it's over. The car stops, but the buzzing of speed and road in their heads does not. Maher finds the door, and his legs, and jogs up under the empty lights of the big Ferris wheel. It's exactly 1:30 a.m. He punches their card into the time clock, flown from New York, and gives the ticket to Roy.

And this, of course, is the end of Roy's Cannonball Run. There are people here—friends and family and a camera crew. The cameraman closes in and asks the questions that you ask: Your thoughts? Why did you do it? And there are jokes and platitudes about Mount Everest and the final frontier, but no real answers.

Why? Because drivers drive. Movies have endings. Records are broken. Perhaps there will be fame, blogs, even an appearance on Conan. Does all that balance against the thousand what-ifs—the nearly cracked axles and the reckless driving, drunk on exhaustion? The crimes that Roy and Maher have committed, state after state, number in the hundreds. There will be months before the statutes of limitation run out, months before this story can finally be published. Roy and Maher have plenty of time to think about what they've done and why.

But for now, the pilot and copilot can only stand with glasses of champagne undrunk. Too tired to know if they are even happy. Or to fully comprehend that their time, 31 hours and 4 minutes coast to coast, has beaten the record by a full hour and three minutes. Or that this record will surely be beaten, again, sometime, by some other drivers, most probably for reasons they won't understand, either.

Cass R. Sunstein

The Polarization of Extremes

*There are six sides to every issue available at the
click of a mouse. So how is it that the internet can
actually limit diversity?*

In 1995 the technology specialist Nicholas Negroponte predicted the emergence of "the Daily Me"—a newspaper that
you design personally, with each component carefully
screened and chosen in advance. For many of us, Negroponte's prediction is coming true. As a result of the internet,
personalization is everywhere. If you want to read essays arguing that climate change is a fraud and a hoax or that the
American economy is about to collapse, the technology is
available to allow you to do exactly that. If you are bored and
upset by the topic of genocide or by recent events in Iraq or
Pakistan, you can avoid those subjects entirely. With just a
few clicks, you can find dozens of Web sites that show you
are quite right to like what you already like and to think
what you already think.

Actually you don't even need to create a Daily Me. With
the internet, it is increasingly easy for others to create one for
you. If people know a little bit about you, they can discover,
and tell you, what "people like you" tend to like—and they
can create a Daily Me, just for you, in a matter of seconds. If

your reading habits suggest that you believe that climate change is a fraud, the process of "collaborative filtering" can be used to find a lot of other material that you are inclined to like. Every year filtering and niche marketing become more sophisticated and refined. Studies show that on Amazon, many purchasers can be divided into "red-state camps" and "blue-state camps," and those who are in one or another camp receive suitable recommendations, ensuring that people will have plenty of materials that cater to, and support, their predilections.

Of course, self-sorting is nothing new. Long before the internet, newspapers and magazines could often be defined in political terms, and many people would flock to those offering congenial points of view. But there is a big difference between a daily newspaper and a Daily Me, and the difference lies in a dramatic increase in the power to fence in and to fence out. Even if they have some kind of political identification, general-interest newspapers and magazines include materials that would not be included in any particular Daily Me; they expose people to topics and points of view that they do not choose in advance. But as a result of the internet, we live increasingly in an era of enclaves and niches—much of it voluntary, much of it produced by those who think they know, and often do know, what we're likely to like. This raises some obvious questions. If people are sorted into enclaves and niches, what will happen to their views? What are the eventual effects on democracy?

To answer these questions, let us put the internet to one side for a moment and explore an experiment conducted in Colorado in 2005, designed to cast light on the consequences of self-sorting. About 60 Americans were brought together and assembled into a number of groups, each consisting of five or six people. Members of each group were asked to deliberate on three of the most controversial issues of the day:

Should states allow same-sex couples to enter into civil unions? Should employers engage in affirmative action by giving a preference to members of traditionally disadvantaged groups? Should the United States sign an international treaty to combat global warming?

As the experiment was designed, the groups consisted of "liberal" and "conservative" enclaves—the former from Boulder, the latter from Colorado Springs. It is widely known that Boulder tends to be liberal and Colorado Springs tends to be conservative. Participants were screened to ensure that they generally conformed to those stereotypes. People were asked to state their opinions anonymously both before and after 15 minutes of group discussion. What was the effect of that discussion?

In almost every case, people held more-extreme positions after they spoke with like-minded others. Discussion made civil unions more popular among liberals and less popular among conservatives. Liberals favored an international treaty to control global warming before discussion; they favored it far more strongly after discussion. Conservatives were neutral on that treaty before discussion, but they strongly opposed it after discussion. Liberals, mildly favorable toward affirmative action before discussion, became strongly favorable toward affirmative action after discussion. Firmly negative about affirmative action before discussion, conservatives became fiercely negative about affirmative action after discussion.

The creation of enclaves of like-minded people had a second effect: It made both liberal groups and conservative groups significantly more homogeneous—and thus squelched diversity. Before people started to talk, many groups displayed a fair amount of internal disagreement on the three issues. The disagreements were greatly reduced as a result of a mere 15-minute discussion. In their anonymous statements, group members showed far more consensus after discussion than before. The discussion greatly widened the rift between liberals and conservatives on all three issues.

The internet makes it exceedingly easy for people to replicate the Colorado experiment online, whether or not that is what they are trying to do. Those who think that affirmative action is a good idea can, and often do, read reams of material that support their view; they can, and often do, exclude any and all material that argues the other way. Those who dislike carbon taxes can find plenty of arguments to that effect. Many liberals jump from one liberal blog to another, and many conservatives restrict their reading to points of view that they find congenial. In short, those who want to find support for what they already think, and to insulate themselves from disturbing topics and contrary points of view, can do that far more easily than they can if they skim through a decent newspaper or weekly newsmagazine.

A key consequence of this kind of self-sorting is what we might call enclave extremism. When people end up in enclaves of like-minded people, they usually move toward a more extreme point in the direction to which the group's members were originally inclined. Enclave extremism is a special case of the broader phenomenon of group polarization, which extends well beyond politics and occurs as groups adopt a more extreme version of whatever view is antecedently favored by their members.

Why do enclaves, on the internet and elsewhere, produce political polarization? The first explanation emphasizes the role of information. Suppose that people who tend to oppose nuclear power are exposed to the views of those who agree with them. It stands to reason that such people will find a disproportionately large number of arguments against nuclear power—and a disproportionately small number of arguments in favor of nuclear power. If people are paying attention to one another, the exchange of information should move people further in opposition to nuclear power. This very process was specifically observed in the Colorado experiment, and in our increasingly enclaved world, it is happening every minute of every day.

The second explanation, involving social comparison, begins with the reasonable suggestion that people want to be perceived favorably by other group members. Once they hear what others believe, they often adjust their positions in the direction of the dominant position. Suppose, for example, that people in an internet discussion group tend to be sharply opposed to the idea of civil unions for same-sex couples and that they also want to seem to be sharply opposed to such unions. If they are speaking with people who are also sharply opposed to these things, they are likely to shift in the direction of even sharper opposition as a result of learning what others think.

The final explanation is the most subtle and probably the most important. The starting point here is that on many issues, most of us are really not sure what we think. Our lack of certainty inclines us toward the middle. Outside of enclaves, moderation is the usual path. Now imagine that people find themselves in enclaves in which they exclusively hear from others who think as they do. As a result, their confidence typically grows, and they become more extreme in their beliefs. Corroboration, in short, reduces tentativeness, and an increase in confidence produces extremism. Enclave extremism is particularly likely to occur on the internet because people can so easily find niches of like-minded types—and discover that their own tentative view is shared by others.

It would be foolish to say, from the mere fact of extreme movements, that people have moved in the wrong direction.

After all, the more extreme tendency might be better rather than worse. Increased extremism, fed by discussions among like-minded people, has helped fuel many movements of great value—including, for example, the civil rights movement, the antislavery movement, the antigenocide movement, the attack on communism in Eastern Europe, and the movement for gender equality. A special advantage of internet enclaves is that they promote the development of positions that would otherwise be invisible, silenced, or squelched in general debate. Even if enclave extremism is at work—perhaps *because* enclave extremism is at work—discussions among like-minded people can provide a wide range of social benefits, not least because they greatly enrich the social "argument pool." The internet can be extremely valuable here.

But there is also a serious danger, which is that people will move to positions that lack merit but are predictable consequences of the particular circumstances of their self-sorting. And it is impossible to say whether those who sort themselves into enclaves of like-minded people will move in a direction that is desirable for society at large or even for the members of each enclave. It is easy to think of examples to the contrary—the rise of Nazism, terrorism, and cults of various sorts. There is a general risk that those who flock together, on the internet or elsewhere, will end up both confident and wrong, simply because they have not been sufficiently exposed to counterarguments. They may even think of their fellow citizens as opponents or adversaries in some kind of "war."

The internet makes it easy for people to create separate communities and niches, and in a free society, much can be said on behalf of both. They can make life a lot more fun; they can reduce loneliness and spur creativity. They can even promote democratic self-government, because enclaves are indispensable for incubating new ideas and perspectives that can strengthen public debate. But it is important to understand that countless editions of the Daily Me can also produce serious problems of mutual suspicion, unjustified rage, and social fragmentation—and that these problems will result from the reliable logic of social interactions.

John Seabrook

Fragmentary Knowledge

*Was the Antikythera Mechanism the world's
first computer?*

In October 2005, a truck pulled up outside the National
Archeological Museum in Athens, and workers began un-
loading an eight-ton X-ray machine that its designer, X-Tek
Systems of Great Britain, had dubbed the Bladerunner.
Standing just inside the National Museum's basement was
Tony Freeth, a 60-year-old British mathematician and filmmaker, watching as workers in white T-shirts wrestled the
Range Rover–size machine through the door and up the
ramp into the museum. Freeth was a member of the Antikythera Mechanism Research Project—a multidiscipli-
nary investigation into some fragments of an ancient mechanical device that were found at the turn of the last
century after 2,000 years in the Aegean Sea and have long
been one of the great mysteries of science.

   Freeth, a tall, taciturn man with a deep, rumbling voice,
had been a mathematician at Bristol University, taking a
PhD in set theory, a branch of mathematical logic. He had
drifted away from the academy, however, and spent most of
his career making films, many of them with scientific themes.
The Antikythera Mechanism, which he had first heard about

some five years earlier, had rekindled his undergraduate love of math and logic and problem solving, and he had all but abandoned his film career in the course of investigating it. He was the latest in a long line of men who have made solving the mystery of the Mechanism their life's work. Another British researcher, Michael Wright, who has studied the Mechanism for more than 20 years, was coincidentally due to arrive in Athens before the Bladerunner had finished its work. But Wright wasn't part of the research project, and his arrival was anticipated with some trepidation.

It had been Freeth's idea to contact X-Tek in the hope of finding a high-resolution, three-dimensional X-ray technology to see inside the fragments of the Mechanism. As it happened, the company was working on a prototype of a CAT scan machine that would use computer tomography to make 3-D X-rays of the blades inside airplane turbines, for safety inspections. Roger Hadland, X-Tek's owner and chief engineer, was interested in Freeth's proposal, and he and his staff developed new technology for the project.

After the lead-lined machine was installed inside the museum, technicians spent another day attaching the peripheral equipment. At last, everything was ready. The first piece to be examined, Fragment D, was placed on the Bladerunner's turntable. It was only about an inch and a half around—much smaller than Fragment A, the largest piece, which measures about six and a half inches across—and it looked like just a small greenish rock or possibly a lump of coral. It was heavily corroded and calcified—the parts of the Mechanism almost indistinguishable from the petrified sea slime that surrounded them. Conservationists couldn't clean off any more of the corroded material without damaging the artifact, and it was hoped that the latest in modern technology would reveal the ancient technology inside.

The Bladerunner began to whirr. As the turntable rotated, an electron gun fired at a tungsten target, which emitted an X-ray beam that passed through the fragment, so that an image was recorded every time the turntable moved a 10th of a degree. A complete 360-degree rotation, resulting in 3,000 images or so, required about an hour. Then the computer required another hour to assemble all the images into a 3-D representation of what the fragment looked like on the inside.

As Freeth waited impatiently for the first images to appear on the Bladerunner's monitor, he was trying not to hope for too much and to place his trust in the skills of the group of academics and technicians who were there with him. Among them, waiting with equal anticipation, were John Seiradakis, a professor of astronomy at the Aristotle University of Thessaloniki; Xenophon Moussas, the director of the Astrophysics Laboratory at the University of Athens; and Yanis Bitsakis, a PhD student in physics. (Mike Edmunds, an astrophysicist at Cardiff University, who was the academic leader of the research project, remained in Wales.) "I was just focused on my relief that this was happening at all, with all the delays of the past four years," Freeth told me. "Honestly, there were times when I thought it would never happen."

One day in the spring of 1900, a party of Greek sponge divers returning from North Africa was forced by a storm to take shelter in the lee of the small island of Antikythera, which lies between Crete and Kythera. After the storm passed, one of the divers, Elias Stadiatis, put on a weighted suit and an airtight helmet that was connected by an air hose to a compressor on the boat and went looking for giant clams, with which to make a feast that evening.

The bottom of the sea dropped sharply, and the diver followed the underwater cliff to a shelf that was about 140 feet below the surface. On the other side of the shelf, an abyss fell away into total darkness. Looking around, Stadiatis saw the remains of an ancient shipwreck. Then he had a terrible shock. There were piles of bodies, all in pieces, covering the ledge. He grabbed one of the pieces before surfacing in order to have proof of what he had seen. It turned out to be a bronze arm.

The following autumn, the sponge divers, now working for the Greek government, returned to the site, and over the next 10 months they brought up many more pieces of sculpture, both marble and bronze, from the wreck, all of which were taken to the National Museum to be cleaned and reassembled. It was the world's first large-scale underwater archeological excavation. Evidence derived from coins, amphorae, and other items of the cargo eventually allowed researchers to fix a date for the shipwreck: around the first half of the first century BC, a time when the glorious civilization of ancient Greece was on the wane, following the Roman conquest of the Greek cities. Coins from Pergamum, a Hellenistic city in what is now Turkey, indicated that the ship had made port nearby. The style of the amphorae strongly suggested that the ship had called at the island of Rhodes, also on the eastern edge of the Hellenistic world and known for its wealth and its industry. Given the reputed corruption of officials in the provinces of the Roman Empire, it is possible that the ship's cargo had been plundered from Greek temples and villas and was on its way to adorn the houses of aristocrats in Rome. The sheer weight of the cargo probably contributed to the ship's destruction.

Most of the marble pieces were blackened and pitted from their long immersion in the salt water, but the bronze sculptures, though badly corroded, were salvageable. Although bronze sculptures were common in ancient Greece, only a tiny number have survived (the bronze was often sold as scrap, melted down, and recast, possibly as weaponry), and most of those have been recovered from shipwrecks. Among the works of art that emerged from the waters near Antikythera are the bronze portrait of a bearded philosopher and the so-called Antikythera Youth, a larger-than-life-size naked young man: a rare specimen of a bronze masterwork, believed to be from the fourth century BC.

Other artifacts included bronze fittings for wooden furniture, pottery, an oil lamp, and item 15087—a shoe box–size lump of bronze, which appeared to have a wooden exterior. Inside were what seemed to be fused metal pieces, but the bronze was so encrusted with barnacles and calcium that it was difficult to tell what it was. With so much early excitement focused on the sculptures, the artifact didn't receive much attention at first. But one day in May 1902, a Greek archeologist named Spyridon Staïs noticed that the wooden exterior had split open, probably as a result of exposure to the air, and that the artifact inside had fallen into several pieces. Looking closely, Staïs saw some inscriptions, in ancient Greek, about two millimeters high, engraved on what looked like a bronze dial. Researchers also noticed precisely cut triangular gear teeth of different sizes. The thing looked like some sort of mechanical clock. But this was impossible, because scientifically precise gearing wasn't believed to have been widely used until the 14th century—1,400 years after the ship went down.

The first analyses of what became known as the Antikythera Mechanism followed two main approaches. The archeologists, led by J. N. Svoronos of the National Museum, thought that the artifact must have been "a kind of astrolabe." A Hellenistic invention, an astrolabe was an astronomical device that was widely known in the Islamic world by the 8th century and in Europe by the early 12th century. Astrolabes were used to tell the time and could also determine latitude with reference to the position of the stars; Muslim sailors often used them, in addition, to calculate prayer times and find the direction of Mecca.

However, other researchers, led by the German philologist Albert Rehm, thought that the Mechanism appeared much too complex to be an astrolabe. Rehm suggested that it might possibly be the legendary Sphere of Archimedes, which Cicero had described in the first century BC as a kind of mechanical planetarium, capable of reproducing the movement of the sun, the moon, and the five planets that could be seen from Earth without a telescope—Mercury, Venus, Mars, Jupiter, and Saturn. Still others, acknowledging the artifact's complexity, thought that it must have come from a much later shipwreck, which may have settled on top of the ancient ship (even though the Mechanism had plainly been crushed under the weight of the ship's other cargo). But, in the absence of any overwhelming evidence one way or the other, until the 1950s the astrolabe theory held sway.

Looking back over the first 50 years of research on the Mechanism, one is struck by the reluctance of modern investigators to credit the ancients with technological skill. The Greeks are thought to have possessed crude wooden gears, which were used to lift heavy building materials, haul up water, and hoist anchors, but historians do not generally credit them with possessing scientifically precise gears—gears cut from metal and arranged into complex "gear trains" capable of carrying motion from one driveshaft to another. Paul Keyser, a software developer at IBM and the author of *Greek Science of the Hellenistic Era,* told me recently, "Those scholars who study the history of science tend to focus on science beginning with Copernicus and Galileo and Harvey, and often go so far as to assert that no such thing existed before." It's almost as if we wished to reserve advanced technological accomplishment exclusively for ourselves. Our civilization, while too late to make the fundamental discoveries that the Greeks made in the sciences—Euclidean geometry, trigonometry, and the law of the lever, to name a few—has excelled at using those discoveries to make machines. These are the product and proof of our unique genius, and we're reluctant to share our glory with previous civilizations.

In fact, there is evidence that earlier civilizations were much more technically adept than we imagine they were. As Peter James and Nick Thorpe point out in *Ancient Inventions,* published in 1994, some ancient civilizations were aware of natural electric phenomena and the invisible powers of magnetism (though neither concept was understood). The Greeks had a tradition of great inventors, beginning with Archimedes of Syracuse (ca. 287–212 BC), who, in addition to his famous planetarium, is believed to have invented a terrible clawed device made up of large hooks, submerged in the sea, and attached by a cable to a terrestrial hoist; the device was capable of lifting the bow of a fully loaded warship into the air and smashing it down on the water—the Greeks reportedly used the weapon during the Roman siege of Syracuse around 212 BC. Philon of Byzantium (who lived around 200 BC) made a spring-driven catapult. Heron of Alexandria (who lived around the first century AD) was the most ingenious inventor of all. He described the basic principles of steam power and is said to have invented a steam-powered device in which escaping steam caused a sphere with two nozzles to rotate. He also made a mechanical slot machine, a water-powered organ, and machinery for temples and theaters, including automatic swinging doors. He is perhaps best remembered for his automatons—simulations of animals and men, cleverly engineered to sing, blow trumpets, and dance, among other lifelike actions.

Although a book by Heron, *Pneumatica,* detailing various of these inventions, has survived, some scholars have dismissed his descriptions as fantasy. They have pointed to the lack of evidence—no trace of any of these marvelous machines has been found. But, as other scholars have pointed out, the lack of archeological evidence isn't really surprising. No doubt, the machines eventually broke down, and, as the know-how faded, there was no one around who could fix them, so they were sold as scrap and recycled. Very few technical drawings or writings remained, because, as Paul Keyser observes, "the texts that survive tend to be the more popular texts—i.e., those that were more often copied—and textbooks, not the research works or the advanced technical ones." Eventually, following the dissolution of the Roman Empire, the technological knowledge possessed by the Greeks disappeared from the West completely.

But, if the Greeks did have greater technological sophistication than we think they did, why didn't they apply it to making more useful things—time- and work-saving machines, for example—instead of elaborate singing automatons? Or is what we consider important about technology—which is, above all, that it is useful—different from what the Greeks considered worthwhile: amusement, enlightenment, delight for its own sake? According to one theory, the Greeks, because they owned slaves, had little incentive to invent labor-saving devices—indeed, they may have found the idea distasteful. Archimedes' claws notwithstanding, there was, as Keyser notes, cultural resistance to making high-tech war machines, because "both the Greeks and the Romans valued individual bravery in war." In any case, in the absence of any obvious practical application for Greek technology, it is easy to believe that it never existed at all.

In 1958, Derek de Solla Price, a fellow at the Institute for Advanced Study in Princeton, went to Athens to examine the Mechanism. Price's interests fell between traditionally defined disciplines. Born in Britain, he trained as a physicist but later switched fields and became the Avalon Professor of the History of Science at Yale; he is credited with founding Scientometrics, a method of measuring and analyzing the pursuit of science. The study of the Mechanism, which incorporates elements of archeology, astronomy, mathematics, philology, classical history, and mechanical engineering, was ideally suited to a polymath like Price, and it consumed the rest of his life.

Price believed that the Mechanism was an ancient "computer," which could be used to calculate astronomical events in the near or distant future: the next full moon, for example. He realized that the inscriptions on the large dial were calendrical markings indicating months, days, and the signs of the zodiac and postulated that there must have been pointers, now missing, that represented the sun and the moon and possibly the planets and that these pointers moved around the dial, indicating the position of the heavenly bodies at different times.

Price set about proving these theories, basing his deductions on the fundamental properties of gearing. Gears work by transmitting power through rotational motion and by realizing mathematical relationships between toothed gear wheels. The Mechanism concentrates on the latter aspect. Price seems to have assumed that the largest gear in the artifact, which is clearly visible in Fragment A, was tied to the movement of the sun—one rotation equaled one solar year. If another gear, representing the moon, was driven by the solar gear, then the ratio of wheels in this gear train must have been designed to match the Greeks' idea of the moon's movements. By counting the number of teeth in each gear, you could calculate the gear ratios, and, by comparing those ratios to astronomical cycles, you can figure out which gears represented which movements.

However, because only a few of the gears appear at the surface of the Mechanism and because many of the gear teeth are missing, Price had to develop methods for estimating total numbers from partial tooth counts. Finally, in 1971, he and a Greek radiographer, Dr. C. Karakalos, were permitted to make the first X-rays of the Mechanism, and these two-dimensional images showed almost all the remaining gear teeth. Price developed a schematic drawing of a hypothetical reconstruction of the internal workings of the Mechanism. In 1974, Price published his research in the form of a 70-page monograph titled *Gears from the Greeks*. He had written, "Nothing like this instrument is preserved elsewhere. Nothing comparable to it is known from any ancient scientific text or literary allusion. On the contrary, from all that we know of science and technology in the Hellenistic Age we should have felt that such a device could not exist."

Price expected his work on the Mechanism to change the history of technology. The Mechanism "requires us to completely rethink our attitudes toward ancient Greek technology," he wrote, and later added, "It must surely rank as one of the greatest mechanical inventions of all time." Price also pointed out that the Mechanism cannot have been the only one of its kind; no technology this sophisticated could have appeared suddenly, fully realized. Not only did the Mechanism demonstrate that our concept of ancient technology was fundamentally incomplete; it also contradicted the neo-Darwinian concept of technical progress in general as a gradual evolution toward ever greater complexity (technological history being the last refuge of the 19th-century belief in progress)—an idea firmly embedded in A. P. Usher's classic 1929 study, *A History of Mechanical Inventions*. As Price writes, it is "a bit frightening to know that just before the fall of their great civilization the ancient Greeks had come so close to our age, not only in their thought, but also in their scientific technology."

But Price's work, though widely reviewed in scholarly journals, did not change the way the history of the ancient world is written. Otto Neugebauer's huge *A History of Ancient Mathematical Astronomy,* which came out the year after *Gears,* relegates the Mechanism to a single footnote. Scholars and historians may have been reluctant to rewrite the history of technology to include research that had lingering doubts attached to it. Also, Price's book was published at the height of the popularity of *Chariots of the Gods,* a 1968 book by the Swiss writer Erich von Däniken, which argued that advanced aliens had seeded the Earth with technology, and Price got associated with UFOs and crop circles and other kinds of fringe thinking. Finally, as Paul Keyser told me, "Classical scholarship is very literary and focuses on texts— such as the writing of Homer, Sophocles, Virgil, or Horace—or it is old-fashioned and historical and focuses on leaders and battles, through the texts of Herodotus and Thucydides, or it is anthropological-archeological and focuses on population distributions and suchlike. So when an archeological discovery about ancient technology arrives, it does not fit, because it's new, it's scientific, and it's not a text. Plus, there is only one such device, and unique items tend to worry scholars and scientists, who quite reasonably prefer patterns and larger collections of data." Whatever the reason, as one scholar, Rob Rice, noted in a paper first presented in 1993, "It is neither facile nor uninstructive to remark that the Antikythera mechanism dropped and sank—twice"— once in the sea and once in scholarship.

The National Museum in Athens took no special pains in displaying the lumps of bronze. Item 15087 wasn't much to look at. When the physicist Richard Feynman visited in 1980, there was little information explaining what the Mechanism was. In a letter to his family, later published in the book *What Do You Care What Other People Think?* the physicist wrote that he found the museum "slightly boring because we have seen so much of that stuff before. Except for one thing: among all those art objects there was one thing so entirely different and strange that it is nearly impossible. It was recovered from the sea in 1900 and is some kind of machine with gear trains, very much like the inside of a modern wind-up alarm clock." When Feynman asked to know more about item 15087, the curators seemed a little disappointed. One said, "Of all the things in the museum, why does he pick out that particular item, what is so special about it?"

For the Greeks, as for other ancient civilizations, astronomy was a vital and practical form of knowledge. The sun and the moon were the basis for calendars by which people marked time. The solar cycle told farmers the best times for sowing and harvesting crops, while the lunar cycle was commonly used as the basis for civic obligations. And, of course, for mariners the stars provided some means of navigating at night.

Xenophon Moussas, one of the two Greek astronomers who are part of the research project, is a compact, soft-spoken man. He grew up in Athens, and as a boy, visiting the museum, he often pondered the Mechanism; now as a professor of astrophysics, he uses it to connect with his undergraduate students, for whom ancient technology is often more compelling than ancient theory.

One evening in January, Moussas led me on a memorable walk around the archeological park in central Athens, which includes both the Greek and the Roman agoras. As a quarter moon shone in the clear night sky, illuminating the ruined temples and markets, Moussas narrated the story of how the ancients slowly learned to recognize patterns and serial events in the movements of the stars and to use them to tell time and to predict future astronomical events. "It was a way of keeping track not of time as we think of it," he told me, "but of the movement of the stars—a deeper time."

For the Greeks, like the Babylonians before them, the year consisted of 12 "lunations," or new-moon-to-new-moon cycles, each of which lasted an average of 29.5 days. The problem with a lunar calendar is that 12 lunar cycles take about 11 days less than one solar cycle. That means that if you don't make regular adjustments to the calendar the seasons soon slip out of synch with the months, and after 18 years or so the summer solstice will occur in December. Finding a system that reconciled the lunar year with the solar year was the great challenge of calendar making.

Most ancient societies readjusted their calendars by adding a 13th, "intercalary" month every three years or so, although methods of calculating the length of these months, and when they should be added, were never precise. Babylonian astronomers hit upon an improvement. They discovered that there are 235 lunar months in 19 years. In other words, if you observe a full moon on April 4, there will be another full moon in that same place on April 4 nineteen years later. This cycle, which eventually came to be known as the Metonic cycle, after the Greek astronomer Meton of Athens, was an extremely useful tool for keeping the lunar calendar and the solar calendar in synch. (The Metonic cycle is still used by the Christian Churches to calculate the correct day for celebrating Easter.) The Babylonians also established what would come to be known as the saros cycle, which is a way of predicting the likely occurrence of eclipses. Babylonian astronomers observed that 18 years, 11 days, and eight hours after an eclipse a nearly identical eclipse will occur. Eclipses were believed by many ancient societies to be omens that, depending on how they were interpreted, could foretell the future of a monarch, for example, or the outcome of a military campaign.

The Greeks, in turn, discovered the Callippic cycle, which consisted of four Metonic cycles minus one day and was an even more precise way to reconcile the cycles of the sun and the moon. But the Greeks' real genius was to work out theories to explain these cycles. In particular, they brought the concept of geometry to Babylonian astronomy. As Alexander Jones, a professor of classics at the University of Toronto, put it to me recently, "The Greeks saw the Babylonian formulas in terms of geometry—they saw all these circles all spinning around each other in the sky. And of course this fits in perfectly with the concept of gearworks—the gears are making little orbits." Some Greek inventor must have realized that it was possible to build a simulation of the movements in the heavens by reproducing the cycles with gears.

But who? Price called the inventor simply "some unknown ingenious mechanic." Others have speculated that the inventor was Hipparchus, the greatest of all ancient Greek astronomers. Hipparchus, who is also believed to have invented trigonometry, lived on the island of Rhodes from about 140 to 120 BC. He detailed a theory to explain the anomalous movements of the moon, which appears to change speed during its orbit of the Earth. Hipparchus is also thought to have founded a school on Rhodes that was maintained after his death by Posidonius, with whom Cicero studied in 79 BC. In one of his letters, Cicero mentions a device "recently constructed by our friend Posidonius," which sounds very like the Mechanism and "which at each revolution reproduces the same motions of the sun, the moon, and the five planets that take place in the heavens every day and night."

As Moussas and I headed uphill toward the Acropolis, he pointed out the spot where Meton's astronomy school and solar observatory had been. On our way back down, we stopped at the famous Tower of the Winds, the now gutted shell of what was the great central clock of ancient Athens. Designed by the renowned astronomer Andronicus of Cyrrhus, it is thought to have been an elaborate water clock on the inside and a sundial on the outside. "But in light of what we know about the Mechanism," Moussas said, "I am beginning to wonder whether this was a much more complicated clock than we think."

When Derek Price died, of a heart attack, in 1983, his work on the Mechanism was unfinished. Although his fundamental insights about the device were sound, he hadn't figured out all the details, nor had he succeeded in producing a working model that was correct in all aspects. That year, in London, a Lebanese man walked into the Science Museum on Exhibition Road with an ancient geared mechanism wrapped in a piece of paper in his pocket. J. V. Field, a curator at the museum, was summoned to examine the artifact, which was in four main fragments. Later, she showed it to Michael Wright, one of the curators of mechanical engineering. According to Wright, the man said that he'd bought the artifact in a street market in Beirut several weeks earlier. The Science Museum eventually bought it from him, and Wright proved that it was a geared sundial calendar that displayed the positions of the sun and the moon in the zodiac. Wright also built a reconstruction of the sundial. The style of lettering on the dial dated the device to the sixth century AD, making it the second oldest geared device ever found, after the Antikythera Mechanism.

In addition to his job as a curator, Wright helped to maintain the old clocks exhibited in the museum. Among them was a replica of the oldest clock that we have a clear account of, constructed in the early 14th century by Richard of Wallingford, the Abbot of Saint Albans. It was a fantastic astronomical device called the Albion ("All-by-One"). Another reconstruction was of a famous planetarium and clock built by Giovanni de' Dondi of Padua in the mid-14th century, known as the Astrarium. Like many students of mechanical history, Wright had noted this odd upwelling of clockwork in Europe, appearing in several places at around the same time. He was familiar with the theory that many of the elements of clockwork were known to the ancients. With the decline of the West, goes this theory, technical expertise passed to the Islamic world, just as many of the Greek texts were translated into Arabic and therefore preserved from loss or destruction. In the 9th century, the Banu Musa brothers, in Baghdad, published the *Book of Ingenious Devices,* which detailed many geared mechanical contrivances, and the 10th-century philosopher and astronomer al-Biruni (973–1048) describes a Box of the Moon—a mechanical lunisolar calendar that used eight gearwheels. The more Wright looked into these old Islamic texts, the more convinced he became that the ancient Greeks' knowledge of gearing had been kept alive in the Islamic world and reintroduced to the West, probably by Arabs in 13th-century Spain.

In the course of this research, Wright became intensely interested in the Antikythera Mechanism. Upon studying Price's account closely, he realized that Price had made several fundamental errors in the gearing. "I could see right away that Price's reconstruction doesn't explain what we can see," he told me. "The man who made the Mechanism made no mistakes. He went straight to what he wanted, in the simplest way possible." Wright resolved to complete Price's work and to build a working model of the Mechanism.

Whereas Price worked mainly on an academic level, approaching the Mechanism from the perspective of mathematical and astronomical theory, Wright drew on his vast practical knowledge of arbors, crown wheels, and other mechanical techniques used in gear-train design. His experience in repairing old grandfather clocks, many of which also have astronomical displays that show the phases of the moon, led him to one of his key insights into the engineering of the Mechanism. He posited that there must have been a revolving ball built in the front dial that indicated the phases of the moon—one hemisphere was black, the other white, and the ball rotated as the moon waxed or waned. Wright also showed how a pin-and-slot construction could be used to model the movement of the moon.

Wright, who is 58, has a British public school demeanor, which is generally courteous and hearty and seemingly rational. But he is prey to dark moods; wild, impolitic outbursts; and overcomplicated personal entanglements— "muddles," he calls them. Although he told me, "I really hate confrontation and antagonism of any kind, even competition," he consistently finds himself in disastrous confrontations with people who should be his allies. Whereas academic researchers are used to collaboration and to sharing resources and insights, Wright is temperamentally more like a lone inventor, working away in secrecy and solitude until he has found the solution.

He did have a collaborator once—Allan Bromley, a lecturer in computer science at the University of Sydney and an expert on Charles Babbage, the 19th-century British mathematician who was the first to conceive of the programmable computer. Bromley used to come to the Science Museum to study Babbage's papers and drawings, and Wright would often lunch with him. In 1990, the pair took new X-rays of the Mechanism, the first since Price's. But Bromley brought the data back to Sydney and would allow Wright to see only small portions of the material. (According to Wright, Bromley confessed "that he had it fixed in his mind that it would be his name, preferably alone, that would be attached to the 'solution.'")

Meanwhile, Wright got into a muddle with his boss at the Science Museum, an "out-and-out bully" who would allow Wright to work on the Mechanism only in his free time. ("We don't do the ancient world," Wright remembers another colleague saying.) This meant that while Wright's wife would go on holiday with their children, Wright would go to the museum in Athens. (Eventually, after years of this routine, he and his wife divorced.)

By the late 1990s, Bromley was dying of cancer. Wright went to see him in Sydney, and Bromley turned much of the data over to him. Just as Wright was finally able to work up their findings for publication, however, he learned of the research project and the effort to take a new set of X-rays of the Mechanism. Instead of viewing this new investigation as a potential boon, he saw it as an improper encroachment on his own turf. "There is a long-established unwritten law concerning the study of Greek antiquities, which is that when one researcher has access to the material, any other researcher is denied access until the first has finished," he wrote to me. "In my case, this understanding was swept aside through the machinations of the group." So, when he arrived at the National Museum while the Bladerunner's X-rays were in progress, he was not excited like the others; he was "angry, tired, and depressed."

The first images of Fragment D to appear on the Bladerunner's monitor were stunning—"so much better than we dared to hope," Freeth told me. "They took your breath way." Inside the corroded rock was what looked like a geared embryo—the incipient bud of an industrial age that remained unborn for a millennium.

Then the team spotted an odd-looking inscription. Andrew Ramsey, X-Tek's computer-tomography specialist, who was operating the viewer, zoomed around inside the 3-D representation until he found the right slice. Written on the side of the gear were the letters *M* and *E—ME*. Was this the maker's mark? Or could *ME* mean "Part 45"? (*ME* is the symbol for 45 in ancient Greek.) Freeth joked that Mike Edmunds had scratched his initials on the fragment. Others suggested that this particular piece of the Mechanism could have been recycled and that the *ME* was left over from some earlier device.

Altogether, the team salvaged about a thousand new letters and inscriptions from the Mechanism—doubling the number available to Price. Together with earlier imaging, the new inscriptions support theories that both Price and Wright had advanced. On Fragment E, for example, the group read "235 divisions on the spiral." "I was amazed," Freeth said. "This completely vindicated Price's idea of the Metonic cycle of 235 lunar months on the upper back dial." They also read words explaining that on the extremity of "the pointer stands a little golden sphere," which probably refers to a representation of the sun on the sun pointer that went around the zodiac dial at the front of the Mechanism. Wright had proposed that the rings of the back dials were made in the form of spirals; the word *eliki,* meaning "spiral," can be seen on Fragment E. On Fragment 22, the number "223" has been observed, pointing to the use of the saros dial as an eclipse indicator.

It was, as Xenophon Moussas put it to me, as if "we had discovered the user's manual, right inside the machine." What had been regarded mainly as an archeological artifact took on a different sort of artifactual status, as an important astronomical text. Very few copies of original astronomical texts remain from the period; most of our knowledge about ancient astronomy comes from other, later astronomers. Little of Hipparchus's writing survives; we rely largely on Ptolemy of Alexandria, who some believe took much of Hipparchus's work and called it his own.

Many of the inscriptions took months to read. Yanis Bitsakis, the PhD student, collaborated with Freeth and the X-Tek team in rendering the X-ray data as computer images, while Agamemnon Tselikas, a leading Greek paleographer, did all the readings and most of the translations. As Bitsakis explained to me, "One of the difficulties in reading the texts was that in ancient Greek there were no spaces between the words, and there are many alternative readings. Also, in many cases the edges of the lines are missing, so we don't know what is continuous text." He and Tselikas would work on the readings through the night, frequently e-mailing and calling other members of the team about new discoveries. Moussas remembers this period, lasting until the spring of 2006, as "the most interesting time in my life." For example, finding the words *sphere* and *cosmos* was extremely moving, Moussas told me: "I felt as though I were communicating with an ancient colleague, through the Mechanism."

One day last month, I paid a visit to Michael Wright, in his book-and-clock-cluttered home in West London. Wright was reading Xenophon, the Greek historian, in ancient Greek. He put the book down and brought out his model of the Mechanism from a cabinet underneath the stairs. In size, it is startlingly similar to a laptop computer, though a bit thicker. On the front dial, in addition to the pointers for the sun and the moon that Price posited, Wright added pointers for the planets and a separate pointer for the day of the year. On the back dial were 223 divisions, marking months in the saros cycle; a similar dial above that showed months in the Metonic cycle. The gears were hidden inside a wooden casing, which had a large wooden knob on one side.

Wright was still a little upset about what he considered the sweeping claims that the research group had made when it published its findings in the November 30, 2006, issue of *Nature*. He almost stayed home from the two-day conference on the Mechanism that the group put on in early December. In the end, he decided to go, taking his wife, Anne, whom he married in 1998, "to stop me from lifting my knee in some chap's groin."

We went upstairs to Wright's workshop. It was filled with tools and pieces of metal, and the air held the pleasantly acrid scent of machine oil. Scattered across the tables and the floor were clever devices that Wright had fashioned out of gears—clocks, astrolabes, engines of various kinds. I recalled Price's description of the maker of the Mechanism— "some unknown ingenious mechanic"—and wondered if this mysterious maker might have been a bit like Wright, with a workshop similarly cluttered with machines.

Wright took his model apart and showed me how all the gears fitted together. I noticed some writing on a rectangular metal plate in the middle of the mechanism, and Wright told me that it was made of recycled bits of brass left over from some previous incarnation.

"So you think that the letters *ME*—"

"Precisely," Wright interjected. "I think they must relate to whatever that bit of metal was used for before."

Then Wright put the machine back together and turned the hand knob that drives the solar gear. It engaged with the smaller gears, through the various gear trains, and the pointers began to spin around the dials. The day-of-the-year pointer moved forward at a regular pace, but the lunar and planetary pointers traced eccentric orbits, sometimes reversing course and going backward, just as the planets occasionally appear to do in the night sky. Meanwhile, the pointers on the back dials crept through the months in the saros and Metonic cycles; eclipses came and went. I noticed that as long as he kept turning the knob Wright himself seemed, for once, perfectly unmuddled.

Until this moment, I had, like many others, continued to puzzle over why, if the Greeks were capable of building such a technically sophisticated device, they used that capacity to construct what is essentially a toy—an intellectual amusement. But as I beheld this whirring, whirling symphony of metal, a perfect simulation of a mechanistic and logical universe, I realized that my notions of practicality were foolish and shortsighted. This machine was much more than a toy; it embodied a whole worldview, and it must have been, for the ancients, wonderfully reassuring to behold.

Julian Dibbell

The Life of the Chinese Gold Farmer

One of China's newest industries—making real
money hunting virtual gold

It was an hour before midnight, three hours into the night
shift with nine more to go. At his workstation in a small,
fluorescent-lighted office space in Nanjing, China, Li Qi-
wen sat shirtless and chain-smoking, gazing purposefully at
the online computer game in front of him. The screen
showed a lightly wooded mountain terrain, studded with
castle ruins and grazing deer, in which warrior monks
milled about. Li, or rather his staff-wielding wizard character, had been slaying the enemy monks since 8:00 p.m.,
mouse-clicking on one corpse after another, each time gathering a few dozen virtual coins—and maybe a magic
weapon or two—into an increasingly laden backpack.

Twelve hours a night, seven nights a week, with only
two or three nights off per month, this is what Li does—for
a living. On this summer night in 2006, the game on his
screen was, as always, World of Warcraft, an online fantasy
title in which players, in the guise of self-created avatars—
night-elf wizards, warrior orcs, and other Tolkienesque
characters—battle their way through the mythical realm of
Azeroth, earning points for every monster slain and rising,

over many months, from the game's lowest level of death-dealing power (1) to the highest (70). More than 8 million people around the world play World of Warcraft—approximately one in every thousand on the planet—and whenever Li is logged on, thousands of other players are, too. They share the game's vast, virtual world with him, converging in its towns to trade their loot or turning up from time to time in Li's own wooded corner of it, looking for enemies to kill and coins to gather. Every World of Warcraft player needs those coins and mostly for one reason: to pay for the virtual gear to fight the monsters to earn the points to reach the next level. And there are only two ways players can get as much of this virtual money as the game requires: they can spend hours collecting it, or they can pay someone real money to do it for them.

At the end of each shift, Li reports the night's haul to his supervisor, and at the end of the week, he, like his nine coworkers, will be paid in full. For every 100 gold coins he gathers, Li makes 10 yuan, or about $1.25, earning an effective wage of 30¢ an hour, more or less. The boss, in turn, receives $3 or more when he sells those same coins to an online retailer, who will sell them to the final customer (an American or European player) for as much as $20. The small commercial space Li and his colleagues work in—two rooms, one for the workers and another for the supervisor—along with a rudimentary workers' dorm, a half hour's bus ride away, are the entire physical plant of this modest $80,000-a-year business. It is estimated that there are thousands of businesses like it all over China, neither owned nor operated by the game companies from which they make their money. Collectively they employ an estimated 100,000 workers, who produce the bulk of all the goods in what has become a $1.8 billion worldwide trade in virtual items. The polite name for these operations is *youxi gongzuoshi,* or gaming workshops, but to gamers throughout the world, they are better known as gold farms. While the internet has produced some strange new job descriptions over the years, it is hard to think of any more surreal than that of the Chinese gold farmer.

The market for massively multiplayer online role-playing games, known as MMOs, is a fast-growing one, with no fewer than 80 current titles and many more under development, all targeted at a player population that totals around 30 million worldwide. World of Warcraft, produced in Irvine, California, by Blizzard Entertainment, is one of the most profitable computer games in history, earning close to $1 billion a year in monthly subscriptions and other revenue. In a typical MMO, as in a classic predigital role-playing game like Dungeons & Dragons, each player leads his fantasy character on a life of combat and adventure that may last for months or even years of play. As has also been true since D&D, however, the romance of this imaginary life stands in sharp contrast to the plodding, mathematical precision with which it proceeds.

Players of MMOs are notoriously obsessive gamers, not infrequently dedicating more time to the make-believe careers of their characters than to their own real jobs. Indeed, it is no mere conceit to say that MMOs are just as much economies as games. In every one of them, there is some form of money, the getting and spending of which invariably demands a lot of attention: in World of Warcraft, it is the generic gold coin; in Korea's popular Lineage II, it is the *adena;* in the Japanese hit Final Fantasy XI, it is called *gil.* And in all of these games, it takes a lot of this virtual local currency to buy the gear and other battle aids a player needs to even contemplate a run at the monsters worth fighting. To get it, players have a range of virtual income-generating activities to choose from: they can collect loot from dead monsters, of course, but they can also make weapons, potions, and similarly useful items to sell to other players or even gather the herbs and hides and other resources that are the crafters' raw materials. Repetitive and time intensive by design, these pursuits and others like them are known collectively as "the grind."

For players lacking time or patience for the grind, there has always been another means of acquiring virtual loot: real money. From the earliest days of MMOs, players have been willing to trade their hard-earned legal tender—dollars, euros, yen, pounds sterling—for the fruits of other players' grinding. And despite strict rules against the practice in the most popular online games, there have always been players willing to sell. The phenomenon of selling virtual goods for real money is called real-money trading, or RMT, and it first flourished in the late 1990s on eBay. MMO players looking to sell their virtual armor, weapons, gold, and other items would post them for auction and then, when all the bids were in and payment was made, arrange with the highest bidder to meet inside the game world and transfer the goods from the seller's account to the buyer's.

Until very recently, in fact, eBay was a major clearinghouse for commodities from every virtual economy known to gaming—from venerable sword-and-sorcery stalwarts EverQuest and Ultima Online to up-and-comers like the Machiavellian space adventure EVE Online and the freeform social sandbox Second Life. That all came to an official end this January, when eBay announced a ban on RMT sales, citing, among other concerns, the customer service issues involved in facilitating transactions that are prohibited by the gaming companies. But by then the market had long since outgrown the tag-sale economics of online auctions. For years now, the vast majority of virtual goods has been brought to retail not by players selling the product of their own gaming but by high-volume online specialty sites like the virtual-money superstores IGE, BroGame, and Massive Online Gaming Sales—multimillion-dollar businesses offering one-stop, one-click shopping and instant delivery of in-game cash. These are the Wal-Marts and Targets of this decidedly gray market, and the same economic logic that leads conventional megaretailers to China in search of cheap toys and textiles takes their virtual counterparts to China's gold farms.

Indeed, on the surface, there is little to distinguish gold farming from toy production or textile manufacture or any of the other industries that have mushroomed across China to feed the desires of the Western consumer. The wages, the margins, the worker housing, the long shifts and endless workweeks—all of these are standard practice. Like many workers in China today, most gold farmers are migrants. Li, for example, came to Nanjing, in the country's industry-heavy coastal region, from less prosperous parts. At 30, he is old for the job and feels it. He says he hopes to marry and start a family, he told me, but doesn't see it happening on his current wages, which are not much better than what he made at his last job, fixing cars. The free company housing means his expenses aren't high—food, cigarettes, bus fare, connection fees at the local *wang ba* (or internet café) where he goes to relax—but even so, Li said, it is difficult to set aside savings. "You can do it," he said, "but you have to economize a lot."

This is the quick-sketch picture of the job, however, and it misses much. To sit at Li's side for an hour or two, amid the dreary, functional surroundings of his workplace, as he navigates the Technicolor fantasy world he earns his living in, is to understand that gold farming isn't just another outsourced job.

When the night shift ends and the sun comes up, Li and

his coworkers know it only by the slivers of daylight that slip in at the edges of the plastic sheeting taped to the windows against the glare. As Li clocks out, another worker takes his seat, takes control of his avatar, and carries on with the same grim routines amid the warrior monks of Azeroth. On most days Li's replacement is 22-year-old Wang Huachen, who has been at this gold farm for a year, ever since he completed his university course in law. Soon, Wang told me, he will take the test for his certificate to practice, but he seems in no particular hurry to.

"I will miss this job," he said. "It can be boring, but I still have sometimes a playful attitude. So I think I will miss this feeling."

Two workstations away, Wang's coworker Zhou Xiaoguang, who is 24, also spends the day shift massacring monks. To watch his face as he plays, you wouldn't guess there was anything like fun involved in this job, and perhaps *fun* isn't exactly the word. As anyone who has spent much time among video gamers knows, the look on a person's face as he or she plays can be a curiously serious one, reflective of the absorbing rigors of many contemporary games. It is hard, in any case, for Zhou to say where the line between work and play falls in a gold farmer's daily routines. "I am here the full 12 hours every day," he told me, offhandedly killing a passing deer with a single crushing blow. "It's not all work. But there's not a big difference between play and work."

I turned to Wang Huachen, who remained intent on manipulating an arsenal of combat spells, and asked again how it was possible that in these circumstances anybody could, as he put it, "have sometimes a playful attitude."

He didn't even look up from his screen. "I cannot explain," he said. "It just feels that way."

In 2001, Edward Castronova, an economist at Indiana University and at the time an EverQuest player, published a paper in which he documented the rate at which his fellow players accumulated virtual goods and then used the current RMT prices of those goods to calculate the total annual wealth generated by all that in-game activity. The figure he arrived at, $135 million, was roughly 25 times the size of EverQuest's RMT market at the time. Updated and more broadly applied, Castronova's results suggest an aggregate gross domestic product for today's virtual economies of anywhere from $7 billion to $12 billion, a range that puts the economic output of the online gamer population in the company of Bolivia's, Albania's, and Nepal's.

Not quite the big time, no, but the implications are bigger, perhaps, than the numbers themselves. Castronova's estimate of EverQuest's GDP showed that online games—even when there is no exchange of actual money—can produce actual wealth. And in doing so Castronova also showed that something curious has happened to the classic economic distinction between play and production: in certain corners of the world, it has melted away. Play has begun to do real work.

This development has not been universally welcomed. In the eyes of many gamers, in fact, real-money trading is essentially a scam—a form of cheating only slightly more refined than, say, offering 20 actual dollars for another player's Boardwalk and Park Place in Monopoly. Some players, and quite a few game designers, see the problem in more systemic terms. Real-money trading harms the game, they argue, because the overheated productivity of gold farms and other profit-seeking operations makes it harder for beginning players to get ahead. Either way, the sense of a certain economic injustice at work breeds resentment. In theory this resentment would be aimed at every link in the RMT chain, from the buyers to the retailers to the gold farm bosses. And, indeed, late last month American WoW players filed a class-action suit against the dominant virtual gold retailer, IGE, the first of its kind.

But as a matter of everyday practice, it is the farmers who catch it in the face. Consider, for example, a typical interlude in the workday of the 21-year-old gold farmer Min Qinghai. Min spends most of his time within the confines of a former manufacturing space 200 miles south of Nanjing in the midsize city of Jinhua. He works two floors below the plywood bunks of the workers' dorm where he sleeps. In two years of 84-hour farming weeks, he has rarely stepped outside for longer than it takes to eat a meal. But he has died more times than he can count. And last September on a warm afternoon, halfway between his lunch and dinner breaks, it was happening again.

The World of Warcraft monsters he faces down—ferocious, gray-furred warriors of the Timbermaw clan of bearmen—are no match for his high-level characters, but they do fight back, and sometimes they get the better of him. And so it appeared they had just done. Distracted from his post for a moment, Min returned to find his hunter-class character at the brink of death, the scene before him a flurry of computer-animated weapon blows. It wasn't until the fight had run its course and the hunter lay dead that Min could make out exactly what had happened. The game's chat window displayed a textual record of the blows landed and the cost to Min in damage points. The record was clear: the monsters hadn't acted alone. In the middle of the fight another player happened by, sneaked up on Min, and brought him down.

Min leaned back and stretched and then set about the tedious business of resurrecting his character, a drawn-out sequence of operations that can put a player out of action for as long as 10 minutes. In farms with daily production quotas, too much time spent dead instead of farming gold can put the worker's job at risk. And in shops where daily wages are tied to daily harvests, every minute lost to death is money taken from the farmer's pocket. But there are times when death is more than just an economic setback for a gold farmer, and this was one of them. As Min returned to his corpse—checking to make sure his attacker wasn't waiting around to fall on him again the moment he resurrected—what hurt more than the death itself was how it happened or, more precisely, what made it happen: another player.

It isn't that WoW players don't frequently kill other players for fun and kill points. They do. But there is usually more to it when the kill in question is a gold farmer. In part because gold farmers' hunting patterns are so repetitive, they are easy to spot, making them ready targets for pent-up anti-RMT hostility, expressed in everything from private sarcastic messages to gratuitous ambushes that can stop a farmer's harvesting in its tracks. In homemade World of Warcraft video clips that circulate on YouTube or GameTrailers, with titles like "Chinese Gold Farmers Must Die" and "Chinese Farmer Extermination," players document their farmer-killing expeditions through that same Timbermaw-ridden patch of WoW in which Min does his farming—a place so popular with farmers that Western players sometimes call it China Town. Nick Yee, an MMO scholar based at Stanford, has noted the unsettling parallels (the recurrence of words like *vermin, rats,* and *extermination*) between contemporary anti-gold-farmer rhetoric and 19th-century U.S. literature on immigrant Chinese laundry workers.

Min's English is not good enough to grasp in all its richness the hatred aimed his way. But he gets the idea. He feels a little embarrassed around regular players and sometimes says he thinks about how he might explain himself to those who believe he has no place among them, if only he could speak their language. "I have this idea in mind that regular players should understand that people do different things in the game," he said. "They are playing. And we are making a living."

It is a distinction that game companies understand all too well. Like the majority of MMO companies, Blizzard has chosen to align itself with the customers who abhor RMT rather than the ones who use it. A year ago, Blizzard announced it had identified and banned more than 50,000 World of Warcraft accounts belonging to farmers. It was the opening salvo in a continuing eradication campaign that has effectively swept millions in farmed gold from the market, sending the exchange rate rocketing from a low of 6¢ per gold coin last spring to a high of 35¢ in January.

Of course, nobody expected the farmers' equally rule-breaking customers to be punished too. Among players, the RMT debate may revolve around questions of fairness, but among game companies, the only question seems to be what is good for business. Cracking down on RMT buyers makes poorer marketing sense than cracking down on sellers, in much the same way that cracking down on illegal drug suppliers is a better political move than cracking down on users. (Only a few companies have found a way to make RMT part of their business model. Sony Online Entertainment, which publishes EverQuest, has started earning respectable revenues from an experimental in-game auction system that charges players a small transaction fee for real-money trades.) As Mark Jacobs, vice president at Electronic Arts and creator of the classic MMO Dark Age of Camelot, put it: "Are you going to get more sympathy from busting 50,000 Chinese farmers or from busting 10,000 Americans that are buying? It's not a racial thing at all. If you bust the buyers, you're busting the guys who are paying to play your game, who you want to keep as customers and who will then go on the forums and say really nasty things about your company and your game."

The cost to farmers of being expelled from WoW can be steep. At the very least, it means a temporary drop in productivity, because the character has to be built up all over again, as well as the loss of all the loot accumulated in that character's account. Given the stakes, some Chinese gold farms have found that the best way to get around their farmers' pursuers is to make it hard to distinguish professionals from players in the first place. One business that specializes in doing just that is located a few blocks from the gold farm where Min Qinghai works. The shop floor is about the same size, with about the same number of computers in the same neat rows, but you can tell just walking through the place that it is a more serious operation. For one thing, there are a lot more workers: typically 25 on the day shift, 25 on the night shift, each crew punching in and out at a time clock just inside the entrance. Nobody works without a shirt here; quite a few, in fact, wear a standard-issue white polo shirt with the company initials on it. There is also a crimson version of the shirt, reserved for management and worn at all times by the shift supervisor, who, when he isn't prowling the floor, sits at his desk before a broad white wall emblazoned with foot-high Chinese characters in red that spell "unity, collaboration, integrity, efficiency."

The name of the business is Donghua Networks, and its specialty is what gamers call "power leveling." Like regular gold farming, power leveling offers customers an end run around the World of Warcraft grind—except that instead of providing money and other items, the power leveler simply does the work for you. Hand over your account name, password, and about $300, and get on with your real life for a while: in a marathon of round-the-clock monster bashing, a team of power levelers will raise your character from the lowest level to the highest, accomplishing in four weeks or less what at a normal rate of play would take at least four months.

For Donghua's owners—26-year-old Fei Jianfeng and 36-year-old Bao Donghua, both former gold-farm wage workers themselves—moving the business out of farming and into leveling was an easy call. Among other advantages, they say, power leveling means fewer banned accounts. Because the only game accounts used are the customers' own, there is much less risk of losing access to the virtual work site. For their workers, however, the advantages are mixed. Though there is a greater variety of quests and quarries to pursue, the pay isn't any better, and some workers chafe at the constraints of playing a stranger's character, preferring the relative autonomy of farming gold.

As one Donghua power leveler said of his old gold-farming job, "I had more room to play for myself."

It may seem strange that a wage-working loot farmer would still care about the freedom to play. But it is not half as strange as the scene that unfolded one evening at nine o'clock in the internet café on the ground floor of the building where Donghua has its offices. Scattered around the stifling, dim wang ba, 10 power levelers just off the day shift were merrily gaming away. Not all of them were playing World of Warcraft. A big, silent lug named Mao sat mesmerized by a very pink-and-purple Japanese schoolgirls' game, in which doe-eyed characters square off in dancing contests with other online players. But the rest had chosen, to a man, to log into their personal World of Warcraft accounts and spend these precious free hours right back where they had spent every other hour of the day: in Azeroth.

Such scenes are not at all unusual. At the end of almost

any working day or night in a Chinese gaming workshop, workers can be found playing the same game they have been playing for the last 12 hours, and to some extent gold-farm operators depend on it. The game is too complex for the bosses to learn it all themselves; they need their workers to be players—to find out all the tricks and shortcuts, to train themselves and to train one another. "When I was a worker," Fan Yangwen, who is now 21 and in Donghua's main office providing technical support, told me, "I loved to play because when I was playing, I was learning." But learning to play or learning to work? I asked. Fan shrugged. "Both."

Fan himself is a striking case of how off-hours play can serve as a kind of unpaid R&D lab for the farming industry. He is that rarest of World of Warcraft obsessives, a Chinese gold farmer who has actually bought farmed gold. ("Sure, I bought 10,000 once," he said. "I don't have time to farm all that!") When Fan shows up at the wang ba after work, it is a minor event; the other Donghua workers pull their chairs over to watch him play—his top-level warlock character is an unbelievable powerhouse that no amount of money, real or virtual, can buy.

What makes Fan's dominance so impressive to his peers is that he achieved it in regions of the game that are all but inaccessible to the working gold farmer or power leveler. Therein lies what is known as the end game, the phase of epic challenges that begins only when the player has accumulated the maximum experience points and can level up no more. The rewards for meeting these challenges are phenomenal: rare weapons and armor pieces loaded with massive power boosts and showy graphics. And the greatest cannot be traded or given away; they can only be acquired by venturing into the game's most difficult dungeons. That requires becoming part of a tightly coordinated "raid" group of as many as 40 other players (any fewer than that, and the entire group will almost certainly "wipe"—or die en masse without killing any monsters of note). Each player has a shot at the best items when they drop, and players must negotiate among themselves for the top prizes. These end-game hurdles have some subtle but significant effects. For one thing, they force the growth of "guilds"—teams of dozens, sometimes hundreds, of players who join together to hit high-end dungeons on a regular basis. For another, they shut farmers out from an entire class of virtual goods—the most marketable in the game if only they could be traded.

For a long time the Donghua bosses, Fei and Bao (known even to employees as Little Bai and Brother Bao), could do no more than nurse their envy of the raiding guilds' access to the end game. But Fan's prowess pointed to another way of looking at it: raiding guilds weren't the competition, they realized; they were the solution. Donghua would put together a team of 40 employees. They would train the team in all the hardest dungeons. And then, for a few hundred dollars, the team would escort any customer into the dungeon of his or her choice. And when the customer's longed-for item dropped, the team would stand aside and let the customer take it, no questions asked. Thus would the supposedly unmarketable end-game treasures find their way into the RMT market. And thus would gold farming, of a sort, find its way at last into the end game.

When Brother Bao and Little Bai put their team together in April of last year, Min Qinghai, a veteran Donghua employee at the time, was among the first to make the roster.

"Before I joined the raiding team, I'd never worked together with so many people," Min told me. They were 40 young men in three adjoining office spaces, and it was chaotic at first. Two or three supervisors moved among them, calling out orders like generals. A dungeon raid is always a puzzle: figuring out which tactics to use to kill each boss is the main challenge; doing so while coordinating 40 players can be dizzying. But members of the team raided just as diligently as they had power-leveled: 12 hours a day, 7 days a week, making their way through the complexities of a different dungeon every day.

There was a lot of shouting involved, at least in the beginning. Besides the orders called out by the supervisors, there were loud attempts at coordination among the team members themselves. "But then we developed a sense of cooperation, and the shouting grew rarer," Min said. "By the end, nothing needed to be said." They moved through the dungeons in silent harmony, 40 intricately interdependent players, each the master of his part. For every fight in every dungeon, the hunters knew without asking exactly when to shoot and at what range; the priests had their healing spells down to a rhythm; wizards knew just how much damage to put in their combat spells.

And Min's role? The translator struggled for a moment to find the word in English, and when I hazarded a guess, Min turned directly to me and repeated it, the only English I ever heard him speak. "Tank," he said, breaking into a rare, slow smile, and why wouldn't he? The tank—the heavily armored warrior character who holds the attention of the most powerful enemy in the fight, taking all its blows—is the linchpin of any raid. If the tank dies, everybody else will soon die, too, as a rule.

"Working together, playing together, it felt nice," Min said. "Very . . . *shuang.*" The word means "open, clear, exhilarating." "You would go in, knowing that you were fighting the bosses that all the guilds in the world dream of fighting; there was a sense of achievement."

The end arrived without warning. One day word came down from the bosses that the 40-man raids were suspended indefinitely for lack of customers. In the meantime, team members would go back to gold farming, gathering loot in five-man dungeons that once might have thrilled Min but now presented no challenge whatsoever. "We no longer went to fight the big boss monsters," Min said. "We were ordered to stay in one place doing the same thing again and again. Every day I was looking at the same thing. I could not stand it."

Min quit and took the farming job he works at still. The new job, with its rote Timbermaw whacking, could hardly be less exciting. But it is more relaxed than Donghua was, less wearying—"Working 12 hours there was like working 24 here"—and he couldn't have stayed on in any case, surrounded by reminders of the broken promise of tanking for what might have been the greatest guild on Earth.

In the meantime, Min is doing his best to forget that his work has anything at all to do with play or that he ever let himself believe otherwise. But even with a job as monotonous as this one, it isn't easy. On his usual hunt one day, he accidentally backed into combat with a higher-level monster. Losing life fast, he grabbed his mouse and started to flee. He hunched over his keyboard, leaning into his flight, flushed now by the chase. His boss, 26-year-old Liu Haibin, an inveterate gamer himself, wandered by and began to cheer him on: "Yeah, yeah, yeah . . . go!"

Finally the monster quit the chase, and Min got away with no consequence more untoward than having to explain himself. "It's instinctual—you can't help it," he said. "You want to play."

How Your Creepy Ex-Co-Workers
Will Kill Facebook

By making it easy for you to be found by people
you're looking to avoid, Facebook and other social
networks are destined to self-destruct.

Facebook's "platform" strategy has sparked much online
debate and controversy. No one wants to see a return to the
miserable days of walled gardens, when you couldn't send a
message to an AOL subscriber unless you, too, were a subscriber and when the only services that made it were the
ones that AOL management approved. Those of us on the
"real" internet regarded AOL with a species of superstitious
dread, a hive of clueless noobs waiting to swamp our
beloved Usenet with dumb flamewars (we fiercely guarded
our erudite flamewars as being of a palpably superior grade),
the wellspring of an endless geyser of free floppy disks and
CDs, the kind of place where the clueless management were
willing and able to—for example—alienate every Vietnamese speaker on earth by banning the use of the word
Phuc (a Vietnamese name) because naughty people might
use it to evade the chatroom censors' blocks on the f-bomb.

Facebook is no paragon of virtue. It bears the hallmarks
of the kind of pump-and-dump service that sees us as sticky,
monetizable eyeballs in need of pimping. The clue is in the

steady stream of e-mails you get from Facebook: "So-and-so has sent you a message." Yeah, what is it? Facebook isn't telling—you have to visit Facebook to find out, generate a banner impression, and read and write your messages using the halt-and-lame Facebook interface, which lags even end-of-lifed e-mail clients like Eudora for composing, reading, filtering, archiving, and searching. E-mails from Facebook aren't helpful messages; they're eyeball bait, intended to send you off to the Facebook site, only to discover that Fred wrote "Hi again!" on your "wall." Like other "social" apps (cough eVite cough), Facebook has all the social graces of a nose-picking, hyperactive six-year-old, standing at the threshold of your attention and chanting, "I know something, I know something, I know something, won't tell you what it is!"

If there was any doubt about Facebook's lack of qualification to displace the internet with a benevolent dictatorship/walled garden, it was removed when Facebook unveiled its new advertising campaign. Now, Facebook will allow its advertisers to use the profile pictures of Facebook users to advertise their products, without permission or compensation. Even if you're the kind of person who likes the sound of a benevolent dictatorship, this clearly isn't one.

Many of my colleagues wonder if Facebook can be redeemed by opening up the platform, letting anyone write any app for the service, easily exporting and importing their data, and so on (this is the kind of thing Google is doing with its OpenSocial Alliance). Perhaps if Facebook takes on some of the characteristics that made the Web work—openness, decentralization, standardization—it will become like the Web itself, but with the added pixie dust of "social," the indefinable characteristic that makes Facebook into pure crack for a significant proportion of internet users.

The debate about redeeming Facebook starts from the

assumption that Facebook is snowballing toward critical mass, the point at which it begins to define "the internet" for a large slice of the world's netizens, growing steadily every day. But I think that this is far from a sure thing. Sure, networks generally follow Metcalfe's Law: "the value of a telecommunications network is proportional to the square of the number of users of the system." This law is best understood through the analogy of the fax machine: a world with one fax machine has no use for faxes, but every time you add a fax, you square the number of possible send/receive combinations (Alice can fax Bob or Carol or Don; Bob can fax Alice, Carol, and Don; Carol can fax Alice, Bob, and Don, etc.).

But Metcalfe's law presumes that creating more communications pathways increases the value of the system, and that's not always true (see Brook's Law: "Adding manpower to a late software project makes it later").

Having watched the rise and fall of SixDegrees, Friendster, and the many other proto-hominids that make up the evolutionary chain leading to Facebook, MySpace, and others, I'm inclined to think that these systems are subject to a Brook's Law parallel: "Adding more users to a social network increases the probability that it will put you in an awkward social circumstance." Perhaps we can call this "boyd's Law" for danah boyd, the social scientist who has studied many of these networks from the inside as a keen-eyed Net anthropologist and who has described the many ways in which social software does violence to sociability in a series of sharp papers. Here's one of boyd's examples, a true story: a young woman, an elementary schoolteacher, joins Friendster after some of her Burning Man buddies send her an invite. All is well until her students sign up and notice that all the friends in her profile are sunburnt, drug-addled technopagans whose own profiles are adorned with digital photos of their painted genitals flapping over the Playa. The teacher inveigles her friends to clean up their profiles, and all is well again until her boss, the school principal, signs up to the service and demands to be added to her friends list. The fact that she doesn't like her boss doesn't really matter: in the social world of Friendster and its progeny, it's perfectly valid to demand to be "friended" in an explicit fashion that most of us left behind in the fourth grade. Now that her boss is on her friends list, our teacher-friend's buddies naturally assume that she is one of the tribe and begin to send her lascivious Friendster grams, inviting her to all sorts of dirty funtimes.

In the real world, we don't articulate our social networks. Imagine how creepy it would be to wander into a coworker's cubicle and discover the wall covered with tiny photos of everyone in the office, ranked by "friend" and "foe," with the top eight friends elevated to a small shrine decorated with Post-it roses and hearts. And yet, there's an undeniable attraction to corralling all your friends and friendly acquaintances, charting them and their relationship to you. Maybe it's evolutionary, some quirk of the neocortex dating from our evolution into social animals who gained advantage by dividing up the work of survival but acquired the tricky job of watching all the other monkeys so as to be sure that everyone was pulling their weight and not napping in the treetops instead of watching for predators, emerging only to eat the fruit the rest of us have foraged.

Keeping track of our social relationships is a serious piece of work that runs a heavy cognitive load. It's natural to seek out some neural prosthesis for assistance in this chore. My fiancée once proposed a "social scheduling" application that would watch your phone and e-mail and IM to figure out who your pals were and give you a little alert if too much time passed without your reaching out to say hello and keep the coals of your relationship aglow. By the time you've reached your 40s, chances are you're out of touch with more friends than you're in touch with: old summer-camp chums, high-school mates, ex-spouses and their families, former co-workers, college roomies, dot-com veterans. Getting all those people back into your life is a full-time job and then some.

You'd think that Facebook would be the perfect tool for handling all this. It's not. For every long-lost chum who reaches out to me on Facebook, there's a guy who beat me up on a weekly basis through the whole seventh grade but now wants to be my buddy; or the crazy person who was fun in college but is now kind of sad; or the creepy ex-co-worker who I'd cross the street to avoid but who now wants to know, "Am I your friend?" yes or no, this instant, please.

It's not just Facebook, and it's not just me. Every "social networking service" has had this problem, and every user I've spoken to has been frustrated by it. I think that's why these services are so volatile: why we're so willing to flee from Friendster and into MySpace's loving arms, from My-Space to Facebook. It's socially awkward to refuse to add someone to your friends list—but removing someone from your friends list is practically a declaration of war. The least awkward way to get back to a friends list with nothing but friends on it is to reboot: create a new identity on a new system and send out some invites (of course, chances are at least one of those invites will go to someone who'll groan and wonder why we're dumb enough to think that we're pals).

That's why I don't worry about Facebook taking over the Net. As more users flock to it, the chances that the person who precipitates your exodus will find you increase. Once that happens, poof, away you go—and Facebook joins SixDegrees, Friendster, and their pals on the scrapheap of Net.history.

Ben Paynter

The Meteor Farmer

Using a souped-up metal detector, a shovel, and a
treasure map, Steve Arnold combs the flat Kansas
wheat fields for rocks from outer space.

For two weeks, Steve Arnold trudged through the dusty
farmland of Kiowa County, Kansas, a six-foot rope trailing
over his shoulder. Tied to the end of the rope was a metal detector cobbled together from PVC pipe and duct tape. Back
and forth Arnold paced, pulling the jury-rigged device
across the dirt, hunting for meteorites. He had already
found a few, but nothing bigger than 100 pounds or so.
Mostly, he found horseshoes. And beer cans. Soon the farmers would want him off their land; planting season was coming. To speed things up, Arnold attached his contraption to
a tractor. He was sure there was a bigger rock out there, just
a few feet beneath the turf.

On a Thursday afternoon, his rig yelped, a shrill beep
sounding through his headphones. He drove forward, tires
pulling in the fine soil, and the detector crescendoed to an
electric wail. Arnold saved the coordinates on his GPS re-
ceiver, marked the spot with a pile of dirt, and pulled out his
cell phone.

Three days later, Arnold and his partner and investor—

an oil and gas attorney from San Antonio named Philip Mani—were attacking the site with a backhoe. After digging down about five feet, Arnold scrabbled into the hole with a shovel and started clearing. Finally, the blade clanged against something metallic. The more dirt he moved, the more meteorite he exposed. They lowered the backhoe scoop and strapped the rock to it. Grinding and whining, the machine pulled free the biggest meteorite Arnold had ever seen.

Its shell was mottled, stippled like ground beef. That's a pattern typical of pallasites, the rarest type of meteorite on Earth. One side was rounded and streamlined by passage through the atmosphere. "It's oriented, Steve!" Mani shouted. "It's oriented!"

About the size of a beer keg, the rock weighed 1,430 pounds, the largest pallasite ever found in the United States. By Arnold's reckoning, it was worth more than $1 million.

When the solar system formed 4.5 billion years ago, the construction debris left behind became what we now call asteroids. As they orbit the sun, asteroids occasionally smash into each other—sometimes with enough force that their guts shatter, the boundary between mantle and core collapsing in a cosmic forge of friction and mass. The result is called a pallasite: iron and nickel folded around the semiprecious gemstone peridot, frozen in mid-alchemy.

Somewhere between 10,000 and 20,000 years ago, according to NASA estimates, one particularly enormous pallasite fell into Earth's gravity well. Eventually, this meteor began plummeting through the atmosphere. The friction and acceleration of entry turned the rock into a glowing fireball, peeling off pieces that trailed behind and lit up the sky over North America.

The chunks hit the ground in southwestern Kansas.

And they sat there, unnoticed, until the latter part of the 19th century, when locals started collecting the heavy, oddly shaped pieces of metal unearthed by plows and wind. Geologists and planetary scientists named the rock collection the Brenham fall, after a nearby town. In 1949, the famous meteorite hunter H. O. Stockwell built a primitive metal-detecting device into a wheelbarrow and found what was, until Steve Arnold showed up, the biggest known lump of Brenham rock, a 1,040-pound behemoth. If you want to see it, take Highway 54 west from Wichita to Greensburg and check out the display in the back of the souvenir shop at the town's only roadside attraction, the Big Well (which is, er, a really big well).

Stockwell's discovery lured other meteorite hunters to the same fallow fields. Locals encountered the rocks all the time. By 2005, people had found several tons of the Brenham meteorite. Collecting pieces used to be the purview of geology nerds, but some of them were discovered by members of an elite clique of globe-trotting meteorite hunters, adventurers paid to retrieve these rare objects. For the past decade, these Indiana Jones types have found a ready market among collectors and art gallery owners, who have come to see meteorites as high-end decorative objects. This new class of hunter means the scientific community gets left out—researchers can't afford the inflated prices of the rocks they want to study.

The genius of Arnold's approach—using homemade maps and souped-up metal detectors—has systematized what was an ad hoc, swashbuckling trade. He finds more specimens, of higher quality, than perhaps any other hunter in the world. And he's doing it all in southwestern Kansas.

Arnold backs his yellow Hummer into the loading dock of Mineral Hunters, a ritzy interior design gallery in Dallas's warehouse district. The name of his new firm, Brenham Meteorite Company, is emblazoned on the truck's tinted windows (motto: Ex Astra, rough Latin for "from the stars"). He says people take him more seriously with the Hummer than they did when he pulled into driveways in a 1986 Dodge van.

Over the years, Arnold has peddled $18,000 worth of meteorites through Mineral Hunters, which showcases and sells his finds. Inside the gallery, Arnold treads carefully— the display room is packed with ornate crystal outcroppings and petrified wood coffee tables, even a 10-foot-tall fossilized dinosaur leg. Arnold is burly enough to worry about knocking something over.

He passes a display featuring black-and-white postcards of himself posing with his 1,430-pound trophy, like a fisherman with a prize catch. Since he unearthed that monster in October 2005, Arnold has found more than two dozen space rocks, a few of which are on display in the gallery. There's a 74-pound specimen with a depression on one side that's cut and polished to reveal its crystal matrix, like amber gems set in a steel mirror. Cool Scoops, marketed as a "Feng Shui personal wealth vase," sells for $39,375. And then there's Braveheart, an 83-pound heart-shaped pallasite with a sparkling orange patina. It's coiled in an elaborate iron stand and has an $88,600 price tag. Jim Penix, the owner of the gallery, comes out to greet Arnold. Braveheart, Penix says, is on hold for a guy who is deciding between buying it or a new Bentley.

A small-timer who barely managed to finish a business administration degree, Arnold got into meteorites 15 years ago. He'd planned to use a metal detector to find buried money left behind by Depression-era ranchers, but while reading archived obituaries for potential targets, he learned that Kansas was a meteorite mother lode.

So Arnold switched rackets, carrying his detector to farmsteads and asking if the residents had any oddly heavy rocks lying around. He bought anything that set off his device. Soon, he fell in with a competitive brotherhood of fireball chasers, guys who raced to fresh fall sites or scavenged former drop zones, chasing a few of the estimated 500 significant meteorites that hit Earth each year. The ranks included Robert "Meteorite Man" Haag, once jailed in Argentina for poaching protected rocks, and Marvin Killgore, a former gold prospector who got Bedouins to help him comb the Sahara desert.

By the mid-1990s, Arnold had found his calling. He scouted the Oman desert by jeep and the Chicago suburbs by bicycle—waving a magnet-tipped broomstick over the ground to pick up the leftover crumbs of a meteorite that hit in 2003.

But chasing falls or finds is tough, hand-to-mouth work. Arnold remembered his early successes in Kansas and wondered: If the ground was full of meteorites, they couldn't have all been found yet, right? He decided to create a sort of treasure map, a way to use the locations of rocks that had already been recovered to find those still buried in the earth.

In 2005, Arnold plotted all the places anyone had found meteorites in Kiowa County. Meteor showers hit the ground in a roughly elliptical pattern called a strewn field, with the bigger chunks at one end. To Arnold, the Brenham-strewn field looked . . . incomplete. He realized that there had to be more big rocks out there.

So he bought a new metal detector. The Pulse Star II sings out a low-pitched tone when the magnetic field it emits hits precious metals; a higher pitch means lead. For most people, that'd be junk, but pallasites have some lead mixed in with the iron and nickel. Arnold would listen for the "junk" signal.

He still needed two more things. First, access to the land. Instead of ringing farmers' doorbells, he turned to his lawyer friend, Mani, who helped him negotiate lease agreements with farmers across Kiowa County. This was the key: In return for a percentage of his profits, they let Arnold search their land for meteorites. He started out with rights to 320 acres.

Second, he needed to amp up his detector. The Pulse Star II's three-foot-square coil wasn't going to cover all that farmland. Arnold bought the largest coil on the market, lashed it to a nine-by-four-foot framework of PVC pipes to cover more surface, and connected the whole thing to the Pulse Star. Then he added plastic wheels and a tow rope to get far enough away that it wouldn't pick up his belt buckle, cell phone, or shovel.

Pretty soon, Arnold was turning up small meteorites. He entered every find onto his map, a recalibration that let him extrapolate the size and rough location of more rocks. He added a longer tether and hitched the detector to a small tractor, and eventually a three-wheeled ATV, to increase his speed.

Even the most jaded of his fellow hunters took notice. Haag, the Meteorite Man, had worked Kiowa County with a handheld detector in the 1980s and walked away with just two finds. "He did it right," Haag says of Arnold, though he insists that the meticulous research and "legal mumbo jumbo" of leasing land would be too sophisticated for him. "I'm glad Steve is doing well with his version," he says. "Everybody does it different. There's gold in them there hills, man! It's treasure, man! It's totally treasure!"

Meteorite hunting wouldn't be so lucrative if it weren't for a music executive named Darryl Pitt. He collected meteorites for years, buying them at rock-bottom rates when the only other buyers were scientists. But in 1995, sitting on a collection numbering in the hundreds, he guessed that people would pay big money for space flotsam. "I needed a mechanism to elevate the profile of my extraterrestrial friends," Pitt says. But he knew he wouldn't get any traction unless he could make people see his "friends'" inherent beauty. A former professional photographer, Pitt started shooting pictures of each of his rocks, lighting them as if they were magazine cover subjects and writing rapturous descriptions in the vein of wine connoisseurship. Then he put them on the block at Phillips International Auctioneers and Valuers, alongside dinosaur eggs and a 3,749-carat opal. The plan worked; the first auction netted close to $200,000.

Pitt is still at it. His online catalog describes meteorites as "objets d'art" with sensuous, zoomorphic shapes—an expert sales job. "Sleek tabletop specimen . . . evocative of the sculpture of Barbara Hepworth," goes one entry. "With a bright platinum patina and compelling from all perspectives."

That auction opened up the market, but it also created a rift between rich collectors and academic researchers. Scientists see meteorites as artifacts from the formation of the solar system or (depending on where the rocks originated) as samples of the geology and atmosphere of other planets. Typically, though, they have to rely on donations to get new research material. "It drives scientists crazy when someone takes a valuable specimen like a Martian meteorite and cuts it up into tiny pieces to make jewelry," says Harry McSween, a geologist at the University of Tennessee and former president of the Meteoritical Society. "Ultimately, you cut them up into such small pieces that you lose the geological information that is contained in the piece, or you have so contaminated them by polishing that you can't make the measurements you want."

Fundamentally, McSween says, commercial concerns don't leave much room for science. "There are issues with how you recover them that certain hunters pay attention to but others don't. Just picking one up can hopelessly contaminate that sample."

Arnold tries to play both sides. He sells his rocks at Mineral Hunters, at auction, on eBay, and on his own Web site. But he also trades with (or sells to) a few scientists who would rather work with hunters than rail against them. The same day as his trip to the gallery, Arnold drove across town to try to make a few trades with Arthur Ehlmann, a geologist who curates the meteorite collection at Texas Christian University. It's one of the biggest academic repositories in the United States; Ehlmann built it from 392 specimens to more than 1,300 by cutting deals with hunters just like Arnold. In general, Ehlmann trades pieces with commercial value for those of scientific interest.

But what about universities unwilling to cut deals? Marvin Killgore—the hunter who worked with the Bedouins—tried to answer that last year, when he became the curator of the University of Arizona's new Southwest Meteorite Center. He announced that he'd buy surplus rock at a fixed rate to create a kind of library for researchers. He would also grade the quality of rocks for private sale in return for a specimen sample of 20 percent or 20 grams, whichever was less. Researchers would get to collect their data, and interior decorators would get to collect their curios. So far, he's received little support from either side of the divide, he says.

Frankly, Arnold is not all that interested in détente, either. The money from private collectors is often too good to pass up. Arnold says he doesn't plan to donate any of his Brenhams—he's even looking for a buyer for the 1,430 pounder, currently on loan to a Wichita, Kansas, science museum. "Patriotism goes a little ways," Arnold says. "But if someone wants it bad enough, then we'll sell it to the person or institution that values it the most."

By summer 2006, Kiowa County was in the grip of a full-fledged meteorite rush. Arnold returned to Greensburg to find farmers armed with handheld detectors searching public roadsides as well as their fields. One resident had created his own metal-detecting rig and was moving from farm to farm, offering royalties that were more generous than Arnold's. And then there were the poachers. Arnold learned that someone was mining without permission on land he was negotiating for when he spotted two Brenhams listed for private auction. (The rocks were eventually returned to the farmer Arnold had been working with.)

Today, the Celestial Museum at the Big Well not only displays Stockwell's original pallasite but also sells shards of Arnold's Brenhams on necklaces and bolo ties. The nearby town of Haviland hosted its first meteorite festival in July. Haviland's mayor wants to build a new space museum and adopt the slogan "The Meteorite Capital of America."

Arnold now owns a small home in Greensburg. Hanging on the wall—the only decoration in the house—is an aerial map of farmland. His original 320-acre lease, just a few miles away, has grown to about 4,000 acres.

Eventually, Kiowa County will be tapped out. Arnold has already accounted for that. He has signed 12 other hunters to join him in using his mapping techniques on eight locations across North America to dig up other valuable meteorites. They were impressed by his success in Kansas and wanted some of the same action. He hasn't told anyone else where they're looking.

Meanwhile, though, there are still pieces of the Brenham fall out there. On an oven-hot day in August, Arnold is in a four-wheeled ATV doing 15 miles per hour through the fields, bouncing over tractor furrows that stand like moguls. His hair is long and shaggy; he's wearing a bright orange T-shirt, cargo shorts, and a pair of Oakley sunglasses. Clouds of dust billow from the detector clattering behind him.

Then a shrill note blares into Arnold's iPod earbuds. The pitch arcs, low to high to low, like a parabola. Because space rocks are fairly symmetrical, they sound the same when approached from any direction. On this plot he's heard mostly uneven signals, all of them what he calls "meteor wrongs": broken plow tips, pieces of horse tack, even the pipeline to a nearby oil pump.

This time the sound goes up and down smoothly, like it's supposed to. "That's a good typical signal," Arnold says. He wands the spot with a weaker handheld detector, but it picks up nothing. Whatever is down there is buried too deep. He grabs a shovel from a rack on the front of the ATV and starts to dig.

At one foot, nothing. At two feet, his handheld beeps faintly. Another foot deeper and the digital screen on the tool identifies the element below: iron. That's a good sign.

Arnold drops to his knees to dig deeper. At four feet the detector wails. Its screen says "Big item. Hot rock." Arnold's face is sweat streaked, caked with mud. He pauses, pulling a magnet across the dirt. It picks up bits of ferrous soil, tinted orange by oxidation.

"Holy Moses," he says. Maybe he should have brought the video camera. He goes back in with the shovel and hits something that thuds. Arnold clears the dirt off the object and reaches into the hole. He yanks hard and comes up with a chunk of rotten barbwire fence post.

"That's man-made something," he says, trying to sound

cheerful as he shovels earth back into the hole. His lease contracts specify that he must refill any hole he digs, and this is today's 10th. Or 11th. But you have to keep a positive attitude when you're on the hunt. "Hey," he says, packing the handheld detector back onto the ATV, "now I know where the meteorites aren't."

Alex Hutchinson

Breaking D-Wave

Has a small business in British Columbia start-up built the world's first viable quantum computer?

If, as Nobel Prize–winning physicist Richard Feynman once said, no one really understands quantum mechanics, then you can appreciate the dilemma that faced Geordie Rose earlier this year as he stood at a podium in front of a room packed with journalists, skeptics, and potential investors, deep in the heart of Silicon Valley. As the chief technology officer and cofounder of D-Wave Systems, a Burnaby-based tech start-up that spun out of the University of British Columbia in 1999, Rose had the daunting task of explaining his company's breakthrough, billed as "the world's first commercial quantum computer." With dark eyebrows looming over a bulldog face and his powerful athlete's body dwarfed by a pair of giant screens flanking the podium he faintly evoked Richard Nixon wilting under the bright lights of the Kennedy debates.

Building a quantum computer—a computer, that is, that harnesses the extraordinarily strange laws of quantum mechanics, which come into play at a subatomic level—has been one of the foremost goals of the scientific world for more than a decade. It would be a fundamentally transformative machine, capable of modeling and predicting the behavior of almost anything in the universe. Most scientists believe it won't be possible to build one for many decades, if ever. Rose begged to differ.

On the twin screens behind him, he called up a giant Sudoku puzzle—"a whimsical example," he acknowledged with a smile. The audience watched as the blanks in the Sudoku were filled in on the screen via a remote connection to the prototype quantum computer, which was back on Burnaby Mountain, housed in a protective copper-walled box at $-273°C$, a hundredth of a degree above absolute zero. While solving a Sudoku is no great accomplishment (for a computer), Rose explained that the puzzle represents a class of mathematical problems that, on larger scales, classical computers are ill-equipped to handle, problems that crop up frequently in business contexts like route planning and database searching.

Pitched to venture capitalists and potential corporate customers, the demo was long on vision and short on quantum mechanics, and the scientific community was underwhelmed; by choosing not to submit their results to peer review before unveiling the computer, D-Wave was skipping the crucial step by which science legitimizes new discoveries. More surprising was the cursory coverage the announcement received in the press. Even without peer-reviewed results, "you would have expected there to be at least some initial spike of excitement," says University of Waterloo researcher Jan Kycia. The media, it appears, simply didn't realize how enormously significant a development a working quantum computer would be, especially one invented by a 75-person Canadian start-up not funded by the U.S. government.

The motto of the United States' National Security Agency (NSA), etched in a plaque across the road from the spy agency's sprawling headquarters off Route 295 in Maryland, is "Always out front." In the world of modern military intelligence, that primarily means staying ahead of rivals in the race for innovative technology to help monitor, decrypt, and analyze surveillance data. This need is the engine that has driven quantum-computing research since 1994, when then AT&T computer scientist Peter Shor made an unexpected discovery. "Shor's algorithm" proved that, in the unlikely event that a quantum computer could be built, it would be able to calculate the factors of very large numbers in a short time. While it's easy to determine that the factors of 21 are 3 and 7, finding the factors of a number with 300 digits would take several millennia for any supercomputer yet built. Since large, impossible-to-factor numbers are used to encrypt everything from secure internet-banking transactions to top-secret government communications, Shor's algorithm had immediate implications for the NSA. Not only is it interested in "reading Osama's e-mail," as some researchers put it, but it has to ensure that encrypted U.S. government communications can't be decoded down the road by yet-to-be-invented technology.

The NSA began pouring money into quantum computing, which up to then had been an obscure idea viewed mainly as a thought experiment, and it was soon joined by other agencies in the Department of Defense. By spreading funding to groups outside the United States, NSA program administrators also kept their fingers on the pulse of progress in virtually every significant research effort in the field. This year, U.S. government spending on quantum-computing research reached $60 million, according to an NSA estimate—a hefty sum for a program whose most concrete progress after more than a decade remains a 2001 experiment that calculated that the factors of 15 are 3 and 5.

Viewed in this light, D-Wave's Sudoku demonstration

looks a little more impressive, especially since Rose says the company has never applied for or received money from the NSA or its sister agencies. D-Wave bills itself as the world's only dedicated commercial quantum-computing enterprise, having raised about $45 million from angel investors and venture capitalists.

Success for D-Wave could be seen as the NSA's worst nightmare: a breakthrough by a privately held foreign company that doesn't disclose its results through the usual scientific channels. It would also run counter to the prevailing wisdom that military funding of basic (knowledge-for-its-own-sake) research is the best means of producing technological innovation. Give money to quantum physicists or astronomers, the argument goes, and your quest for a greater understanding of the universe will produce by-products like lasers and moon landings.

An opposing school of thought, articulated by historian Paul Forman, argues that post–World War II military funding has both diverted scientists from the pursuit of knowledge and proven to be a mostly ineffective way of developing new technology. "Most technological advance is incremental," says Forman, a curator at the Smithsonian's National Museum of American History in Washington, "and incremental technological advance is not, by definition, coming from basic research." D-Wave, in contrast to its NSA-funded counterparts in academia, is trying to build a computer, not make new discoveries about quantum mechanics, an approach that may give them an advantage. "It's the computation part that is important," Rose says, "not the quantum part."

Classical computers encode their data as a series of binary bits, which can take on one of two different values, usually denoted by zero and one. Quantum computers, on the other hand, take advantage of the rules of quantum mechanics, which were developed to explain a series of puzzling experimental results in the early 20th century. It turns out that the rules for governing the motion of very small particles such as electrons are wildly different from the ones we've observed from, say, throwing baseballs around. An electron can be in two places at once, it can "tunnel" through walls, and it can teleport. As a result, a "qubit," the quantum version of a bit, can be zero, one, or—and this is the crucial part—zero and one at the same time. This is where the remarkable power of quantum computers resides, and it increases exponentially if you combine qubits together: a group of just 16 qubits can represent 65,536 different numbers *simultaneously*.

Cracking a code is essentially like trying to guess a number between one and a trillion (the harder the code, the bigger the number). While a classical computer would have to try each number in succession until it found the right one, a quantum computer could try all the numbers at once. This popular explanation of how a quantum computer works, however, isn't quite right. After all, if you ask a trillion questions at once ("Is zero through a trillion the right password?"), the answer you get back ("Yes") isn't very useful. Shor's algorithm does indeed involve manipulating qubits that represent many numbers simultaneously, but it requires a more circuitous search for patterns that enable it to reject the wrong passwords and spit out only the right one.

However you think about it, the result is a computer that can do things that were previously impossible, such as factoring huge numbers quickly. But factoring is far from the only application of quantum computing, though it has dominated the agenda thanks to the NSA's interest and bankroll. "I've always thought that Shor's algorithm has been a very big negative overall for quantum computing, because it's thrown everybody off track," D-Wave's Rose says. "The really valuable applications are not in code breaking—those are not interesting from an industrial perspective, because they don't lead to recurring large markets."

The most radical application for quantum computers is something called quantum simulation. If you wanted to know whether a certain complicated molecule would make a good drug, you could, in theory, solve the quantum-mechanical equations that govern the motion of all the electrons and protons and other components of the molecule, which would tell you exactly how it will behave. But the math is too complex for us to puzzle out. Because qubits follow the same quantum mechanical rules that electrons and protons do, a quantum computer actually embodies those equations and can solve them effortlessly. If the computer is powerful enough, it could model just about anything in the universe. "This is not like going from the Pentium II to the Pentium III," says Ray Laflamme, director of Waterloo's Institute for Quantum Computing, the world's largest dedicated quantum-computing research center. "It's fundamentally different."

The difference is such that no one has really been sure how to build a quantum computer. The qubit has to be something that can physically take on two values representing zero and one, and there are wildly different ideas about what that should be. The computer that factored 15 in 2001 was a 7-qubit system that used nuclear magnetic resonance (the same process used for MRI imaging) to toggle the spins of a test tube of carbon and fluorine atoms. This approach is great for toy-sized computers but seemingly impossible to scale up to the thousands or millions of qubits needed for a useful quantum computer. The opposite is true for superconductor-based systems in which zero and one are sometimes represented by electrical currents flowing either clockwise or counterclockwise around a loop of superconducting metal. Building even a few superconducting qubits has proven to be exceptionally challenging, but if that hurdle is cleared, it should be relatively straightforward to scale the system up using production techniques already developed by the computer industry.

D-Wave's demonstration system is a 16-qubit superconductor-based machine. This is remarkable because, of all the groups around the world working on superconducting qubits, none has succeeded in operating more than 2 qubits together. The problem is that the quantum state that allows zero and one to exist simultaneously is extremely sensitive to outside perturbations. Noise from the outside environment, stray electrical signals, and tiny temperature fluctuations can all cause the qubit to lose its quantum-mechanical properties, a process known as decoherence. For scientists confronted with D-Wave's claims that they've operated a 16-qubit computer, the immediate question is: how did they overcome decoherence? "If you go and ask anybody in the world who's leading in putting qubits together, the first thing you'll ask them is, 'What is the decoherence time?'" says Laflamme. "And if they say 'I don't know,' you'll be very skeptical of the rest of the thing."

D-Wave has certainly met with skepticism, but Rose, who won three Canadian wrestling titles before starting his PhD in theoretical physics at UBC, seems to relish the fight. D-Wave, he explains, has adopted a modified quantum-computer architecture that, while not capable of executing Shor's algorithm, has the advantage of being less sensitive to decoherence and can still tackle a number of other commercially interesting algorithms. He insists that the emphasis on the computer's building blocks is a diversion from the real test of a quantum computer, which is how much time it needs to tackle problems of various sizes. "That's really the only smoking gun for quantum computation," he says. The current 16-qubit system is too small to test performance on large problems, but D-Wave's ambitious road map calls for a 32-qubit system by the end of this year and 1,024 qubits by the third quarter of 2008. While scientists are clamoring for evidence that D-Wave's computer is truly quantum, as far as Rose is concerned, the question will be answered purely on the basis of the computer's performance within the next year. "And you know, it could turn out that the whole thing won't work," he says. "That's always a risk. But it won't be for lack of effort, I'll tell you that much."

When I finished my physics PhD and headed out on the postdoctoral job-interview circuit in late 2001, interest in quantum computing was exploding. I added a couple of slides to the end of my standard presentation, drawing an extremely tenuous link between my thesis work and quantum computing, and even went so far as to insert "quantum computing" in the title of a talk I gave at a university in the Netherlands. I finally accepted a position in the quantum-computing group at an NSA lab in Maryland, where, despite the name, the research we did had little connection to the world of qubits. Instead, we were exploring the boundaries between the classical and quantum worlds. We know, for instance, that an atom can be in two places at once and a baseball can't. But what about 10 atoms? Or 1,000 atoms? Where is the dividing line between atoms and baseballs? You could argue—as we did to our funders—that it would be useful to know this in order to build a quantum computer, but, really, our primary interest was trying to answer one of the great questions of modern physics.

The duplicitous game of justifying fundamental science by promising that it will lead to a magic computing machine or some equivalent is, increasingly, a fact of life for scientists in every field. Forman, the Smithsonian historian, argues that the transition from modernity to postmodernity three decades ago was marked by a dramatic flip in the status of science and technology. Until then, "science was very closely connected with knowledge for its own sake," he says. Now, the process of scientific inquiry is only justified by its ends: useful gadgets. "Our postmodern perspective really does put the very existence of science into question." For those of us who view understanding our universe as a worthy goal, we can be grudgingly grateful that the military's approach to technology development still leaves room, however constrained, for basic science.

There is also room for D-Wave's purely technological, engineering-style approach. Their glossy demonstration in February may have left some scientists fuming, but it spurred plenty of ideas for applications from potential industry partners, Rose says. As the company forges ahead, intent on doubling and redoubling the number of qubits in its system rather than painstakingly ensuring that the qubits it already has are really behaving as they should, scientists will continue to be irritated. But what D-Wave is trying to do shouldn't be confused with science: it is simply trying to build a computer, using technology that may or may not be ready for the task. In that spirit, it's worth remembering Rose's words as he stood at the podium demonstrating a final sample application: the famously difficult mathematical problem of finding the optimal seating plan for a wedding. The list of guests and their seating requirements was sent from the California auditorium to the system in British Columbia, which churned out the answer. As the giant screen displayed a solution that separated two guests who wanted to sit together, Rose turned to face the audience, eyes blinking innocently, and said, "Sometimes you just can't satisfy everybody."

Thomas Goetz

Your DNA Decoded

23andMe Will Decode Your DNA for $1,000.
Welcome to the Age of Genomics.

At the age of 65, my grandfather, the manager of a leather tannery in Fond Du Lac, Wisconsin, suffered a severe heart attack. He had chest pains and was rushed to the hospital. But that was in 1945, before open heart surgery, and he died a few hours later. By the time my father reached 65, he was watching his diet and exercising regularly. That regimen seemed fine until a couple of years later, when he developed chest pains during exercise, a symptom of severe arteriolosclerosis. A checkup revealed that his blood vessels were clogged with arterial plaque. Within two days he had a triple bypass. Fifteen years later (15 years that he considers a gift), he's had no heart trouble to speak of.

I won't reach 65 till 2033, but I have long assumed that, as regards heart disease, my time will come. My genes have predetermined it. To avoid my father's surgery or my grandfather's fate, I try to eat healthier than most, exercise more than most, and never even consider smoking. This, I figure, is what it will take for me to live past 65.

Turns out that my odds are better than I thought. My

DNA isn't pushing me toward heart disease—it's pulling me away. There are established genetic variations that researchers associate with a higher risk for a heart attack, and my genome doesn't have any of those negative mutations; it has positive mutations that actually reduce my risk. Like any American, I still have a good chance of eventually developing heart disease. But when it comes to an inherited risk, I take after my mother, not my father.

Reading your genomic profile—learning your predispositions for various diseases, odd traits, and a talent or two—is something like going to a phantasmagorical family reunion. First you're introduced to the grandfather who died 23 years before you were born and then you move along for a chat with your parents, who are uncharacteristically willing to talk about their health—Dad's prostate, Mom's digestive tract. Next, you have the odd experience of getting acquainted with future versions of yourself, 10, 20, and 30 years down the road. Finally, you face the prospect of telling your children—in my case, my eight-month-old son—that he, like me, may face an increased genetic risk for glaucoma.

The experience is simultaneously unsettling, illuminating, and empowering. And now it's something anyone can have for about $1,000. This winter marks the birth of a new industry: Companies will take a sample of your DNA, scan it, and tell you about your genetic future, as well as your ancestral past. A much anticipated Silicon Valley start-up called 23andMe offers a thorough tour of your genealogy, tracing your DNA back through the eons. Sign up members of your family and you can track generations of inheritance for traits like athletic endurance or bitter-taste blindness. The company will also tell you which diseases and conditions are associated with your genes—from colorectal cancer to lactose intolerance—giving you the ability to take preventive action. A second company, called Navigenics, focuses on matching your genes to current medical research, calculating your genetic risk for a range of diseases.

The advent of retail genomics will make a once-rare experience commonplace. Simply by spitting into a vial, customers of these companies will become early adopters of personalized medicine. We will not live according to what has happened to us (that knee injury from high school or those 20 pounds we've gained since college) nor according to what happens to most Americans (the one-in-three chance men have of getting cancer or women have of dying from heart disease or anyone has for obesity). We will live according to what our own specific genetic risks predispose us toward.

This new industry draws on science that is just beginning to emerge. Genomics is in its earliest days: The Human Genome Project, the landmark effort to sequence the DNA of our species, was completed in 2003, and the research built on that milestone is only now being published. The fact that any consumer with $1,000 can now capitalize on this project is a rare case of groundbreaking science overlapping with an eager marketplace. For the moment, 23andMe and Navigenics offer genotyping: the strategic scanning of your DNA for several hundred thousand of the telltale variations that make one human different from the next. But in a few years, as the price of sequencing the entire genome drops below $1,000, all 6 billion points of your genetic code will be opened to scrutiny.

To act on this data, we first need to understand it. That means the companies must translate the demanding argot of genetics—alleles and phenotypes and centromeres—into something approachable, even simple, for physicians and laypersons alike. It's one thing for a doctor to tell patients that smoking is bad for them or that their cholesterol count is high. But how are you supposed to react when you're told you have a genetic variation at rs6983267, which has been associated with a 20 percent higher risk of colorectal cancer? And what are physicians, most likely untrained in and unprepared for genomic medicine, to do when a patient comes in wielding a printout that indicates a particular variation of a particular gene?

This new age of genomics comes with great opportunity—but also great quandaries. In the genomic age, we will no longer have the problem of not knowing, but we will face the burden of whether we want to know in the first place. We'll learn what might be best for us in life and then have to reckon with the risk and perhaps the guilt of not acting on that knowledge. We will, counterintuitively, face even more pressure to conduct our lives carefully, strictly, and cautiously; we'll practice the art of predictive diagnosis and receive a demanding roster of things to avoid, things to do, and treatments to receive—long before there's any physical evidence of disease. And, yes, we will know whether our children are predisposed to certain traits or talents—athletics or music or languages—and encourage them to pursue certain paths. In short, life will become a little more like a game of strategy, where we're always playing the percentages, trying to optimize our outcomes. "These are enormously large calculations," says Leroy Hood, a pioneer of genomic sequencing and cofounder of the Institute for Systems Biology in Seattle, who suggests that if we pay attention and get the math right, "it's not a stretch to say that we could increase our productive life spans by at least a decade."

The question was surely strange. In February 2005, Anne Wojcicki sat down at the so-called Billionaires' Dinner, an annual event held in Monterey, California, and asked her tablemates about their urine. She was curious whether, after eating asparagus, they could smell it when they urinated.

Among those at her table were geneticist Craig Venter; Ryan Phelan, the CEO of DNA Direct, a San Francisco genetic-testing company; and Wojcicki's then boyfriend (and now husband), Sergey Brin, cofounder of Google. Most could pick up the smell of methyl mercaptan, a sulfur compound released as our guts digest the vegetable. But some had no idea what Wojcicki was talking about. They had, it seems, a genetic variation that made the particular smell imperceptible to them.

Soon, the conversation turned to a growing problem: While researchers are amassing great knowledge about certain genes and genetic variations, there is no way for people to access that data for insights about themselves and their families—to Google their genome, as it were. As a biotech and health care analyst at Passport Capital, a San Francisco hedge fund firm, Wojcicki knew that the pharmaceutical industry was already at work on tailoring drugs to specific genetic profiles. But she was intrigued by the prospect of a database that would compile the available research into a single resource.

Linda Avey wasn't at the dinner, but she wished she had been when she read about it later that year in David Vise and Mark Malseed's book, *The Google Story*. At the time, Avey was an executive at Affymetrix, the company that had pioneered some of the tools for modern genetic research. For nearly a year, she had been mulling the idea of a genotyping tool for consumers, one that would let them plumb their own genome as well as create a novel data pool for researchers. She even had a placeholder name for it: Newco. "All the pieces were there," Avey says. "All we needed was the money, as usual, and computational power." Two things that Google has plenty of. Around the time she read Vise and Malseed's book, Avey had a dinner scheduled with a Google executive. She asked Wojcicki to join them, and the two quickly hit it off. Within a few months, they had settled on the idea behind 23andMe: Give people a look at their genome and help them make sense of it. (The company's name is a reference to the 23 pairs of chromosomes that contain our DNA.)

Brin offered to be an angel investor. "Sergey was like, 'Come up with something in three months and launch it,'" Wojcicki says. "We thought it would be so fast." In fact, the project took more than 18 months from conception to launch. Last spring, Google invested $3.9 million in 23andMe (part of the proceeds repaid Brin, who has since recused himself from the investment). The company, which now has more than 30 employees in a building down the road from Google, feels very much like the quintessential start-up. In the entry hall, alongside two Segways (a gift from inventor Dean Kamen), stands a herd of pedal-pusher bicycles. On a whiteboard in the hall, someone has scrawled an anxiety meter. Current threat level: slight deformation (engineering-speak for moderate stress). But that level had been crossed out and the alert upped to bananas.

Still, 23andMe is hardly a typical Valley outfit. Instead of widgets and Ajax apps, the cubicle chatter more likely concerns Klinefelter's syndrome and hermaphrodites. Such banter underscores a major challenge for the company: making customers comfortable with the strange vocabulary and discomfiting implications of genetics. As Avey notes, when you're asking your customers for their spit, best to have an especially strong relationship.

A lot of spit, as it turns out. It takes about 10 minutes of slavering to fill the 2.5-milliliter vial that comes in the fancy lime box provided by 23andMe. Wrap it up, call FedEx, and two to four weeks later you get an e-mail inviting you to log in and review your results. There are three main sections to the Web site: Genome Labs, where users can navigate through the raw catalog of their 23 pairs of chromosomes; Gene Journals, where the company correlates your genome with current research on a dozen or so diseases and conditions, from type 2 diabetes to Crohn's disease; and Ancestry, where customers can reach back through their DNA and discover their lineage, as well as explore their relationships with ethnic groups around the world. Family members can share profiles, trace the origin of particular traits, and compare one cousin's genome to another in a fascinating display of DNA networking. Avey herself has had roughly 30 members of her extended family genotyped, spanning four generations. The effort has turned her clan into what is likely the most thoroughly documented gene pool in the world.

It's the Gene Journals, though, that could really change people's lives. Here customers learn their personalized risk for a particular condition, calculated according to whether their genotype contains markers that research has associated with specific risks. Wojcicki stresses, though, that 23andMe's results are not a diagnosis. "It's simply your information," she insists. In part, this distinction is to make sure the company doesn't run afoul of the Food and Drug Administration, which strictly regulates diagnostic testing for disease but has been slow to respond to the more transformational aspects of genomics. But the caveat also matters because the influence of genetics varies from disease to disease; some conditions have a strong heritable component, while others are determined more by environmental factors.

With its emphasis on disease risks, Navigenics is more comfortable offering something closer to a diagnosis. "If I tell you you've got a genetic likelihood of getting colon cancer, you're going to get a colonoscopy early," says Navigenics cofounder David Agus, a prominent oncologist and director of the Spielberg Family Center for Applied Proteomics at

Cedars-Sinai Medical Center in Los Angeles. "And that's going to save lives."

Both companies draw a good lesson from the bad example of the body scan industry. When storefront CT scanning machines popped up in the late 1990s, the idea seemed golden to many radiologists and entrepreneurs: Customers could go directly to an imaging center and get an early look at possible tumors or polyps for about $1,000. But the market cratered by 2005, when it became evident that insurers wouldn't pay for the scan without a prior diagnosis and customers wouldn't pay out of pocket for frequent scans. What's more, the false-positive rate was jarringly high, and anxious customers often raced back to their doctor with an image showing, for instance, benign kidney or liver cysts, only to be told that they were harmless incidental lumps.

In other words, there was too much noise and not enough signal. So both 23andMe and Navigenics are determined not to simply shovel along raw research, with scary one-off results indistinguishable from well-established correlations. In-house experts at both companies have filtered and vetted hundreds of studies; only a handful are deemed strong enough to incorporate into their library of conditions, which is used for personalized risk calculations. The hope is that this will reduce or eliminate false alarms and let customers trust the experience—maybe even enjoy it.

One afternoon I was working up my own 2.5 milliliters of spit at the company's office when Jimmy Buffett dropped by to get an early peek at his results. A few months earlier, the singer had let 23andMe peruse his genotype and compare his genealogy to Warren Buffett's. The two men had long wondered if they were somehow related (they aren't, it turns out). Now Jimmy wanted to check out the whole experience. He sat down in front of a laptop in Wojcicki's office, and she looked over his shoulder, guiding him through the site. First he clicked through his ancestral genome, noting that his maternal lineage showed a strong connection to the British Isles. "So the women came over with the Saxon invasion; pretty cool," he said. Another click and he perused his similarity to other ethnic groups, spotting a strong link to the Basque region of Spain. "No wonder I like Basque food so much," he noted.

Then he clicked over to see his disease risks—and was transfixed. "Wow. Right, that's about right for my family," he said as he ran through various conditions. After about 45 minutes of self-discovery, he leaned back in his chair to put it all together. "Boy, this can get pretty fascinating. And every time some research comes out, I can log on and see how it works for me. I get it," Buffett said with a laugh. "You guys are mad scientists."

Gregor Mendel began growing peas in his abbey garden in the 1850s, just a simple monk curious about the differences among the plants. A member of the Augustinian order, Mendel took to his garden experiments with characteristic discipline and rigor. He grew some peas with green seeds and others with yellow ones, some with violet flowers and others with white, some with round seed pods and others with wrinkled pods, and so on—at least 10,000 plants in all. By the time he was done, he had established the principles of genetic inheritance, identifying some traits as dominant and others as recessive. (Less celebrated is his later work breeding honeybees; though his hybridized African and South American bees produced wonderful honey, they were exceptionally vicious, and he destroyed them.)

More than a century later, Mendel's basic concepts remain the cornerstone of genetics. We now understand his traits as genes, and genes as sections of DNA—a strand of 3 billion pairs of ATGC (adenine and thymine, and guanine and cytosine), the nucleotides that compose our genome.

Since 1983, when the gene associated with Huntington's disease was first linked to a particular chromosome, most genetic discoveries have worked like Mendel's peas: They have focused on traits associated with single genes. These so-called monogenic conditions—diseases like hemochromatosis (where the body absorbs too much iron) or Huntington's disease—are easy to research, because the associations are pretty much binary. If you have the genetic mutation, you're almost certain to develop the disease. That makes them easy to screen for, too. There are now tests for more than 1,400 of these diseases: prenatal screening for cystic fibrosis, mutations in BRCA1 and BRCA2 genes that convey a strong risk for breast cancer, and so forth. This is the sort of genetic testing most of us are familiar with. And such screening can be extremely useful. Careful testing for Tay-Sachs disease among Ashkenazi Jews, for instance, has led to a 90 percent reduction in the disease in the United States and Canada.

But as genetic research has progressed, the idea that most diseases will have a clearly defined, single genetic component—what's known as the "common disease, common gene" hypothesis—has turned out to be mostly wishful thinking. In fact, the 1,400 conditions that are currently tested for represent about 5 percent of diseases in developed countries, meaning that for 95 percent of diseases there's something more complicated going on.

Most conditions, it turns out, develop from a subtle interplay among several genes. They are said to be multigenic, not monogenic. And while scientists have made progress connecting the deterministic dots between rare genes and rare conditions, they face a far greater challenge understanding the subtler genetic factors for those more common conditions that have the major impact on society. "We're learning plenty about the molecular basis of disease—that's the revolution right now," says Eric Lander, founding director of the Broad Institute and one of the leaders of the Human Genome Project. "But whether that knowledge translates into personalized predictions and personalized therapeutics is unknown." In other words, not all genes are as simple to understand as Mendel's peas.

The source of this complexity lies in our SNPs, or single nucleotide polymorphisms, the single-letter mutations among the base pairs of DNA—swapping an A for a G or a T for a C—that largely determine how one human is genetically different from another. Throughout our 6 billion bits of genetic code, there are millions of SNPs (pronounced "snips"), and some untold number of those play a role in our predilection for disease. For researchers like Lander, the main challenge is establishing which SNPs—or which constellation of SNPs—affect which conditions.

Consider, for instance, the many ways that a human heart can go bad. The arteries supplying blood to the heart can be clogged with plaque, constricting blood flow until the organ goes into arrest. Or a valve in the heart can leak, spilling blood into the lungs and causing pulmonary edema. Or the tissues of the organ itself can be weakened, as in cardiomyopathy, so that the muscle fails to pump enough blood throughout the body. Each of these conditions has specific terminology, causes, and treatments, but they are all versions of heart disease, which is the leading killer in the United States. And each condition may have its own genetic component, or be influenced by a range of genetic components, with each case of the illness a unique combination of genetic variables and environmental factors. So establishing the genetic component of heart disease means, in actuality, accounting for a daunting variety of conditions and tracking the influence of a broad number of genetic variations, as well as separating them from environmental components.

Now, thanks to a series of complementary innovations, geneticists have begun teasing apart the complexity. First, the Human Genome Project, completed in 2003, provided a map for our common genomic sequence. Next, 2005 saw the completion of the first phase of the International HapMap Project, a less-celebrated but equally ambitious effort that cataloged common patterns of genetic variations, or haplotypes, SNP by SNP. That helped researchers know where they should focus their attention. And, finally, by mid-2006 the price of genotyping microarrays—the matchbox-sized chips that can detect SNP variations from genome to genome—had dropped to a level that let scientists greatly increase the pace and scope of their research.

As these three factors have converged, the pace of discovery has taken off, producing a startling number of new associations between SNPs and disease. Even the sober *New England Journal of Medicine* described trying to keep up with the research as "drinking from the fire hose." Lander calls it a 20-year dream coming to fruition. "2007 has been one of those magical years where the entire picture comes into focus. Suddenly we have the tools to apply to any problem: cancer, diabetes—a huge list of diseases. It's just a stunning explosion of data. Pick a metaphor: We've now landed on this new continent, and the people are out there exploring it, and we're finding mountains and waterfalls and rivers. We're turning on lights in dark rooms. We're finding pieces to the jigsaw puzzle."

Clearly, this is an exciting time to be a geneticist. And, it turns out, a consumer, too.

Come late September, Avey and Wojcicki invited their board of scientific advisers to Mountain View, California, for one last review of the site before launch. The meeting began around noon. Avey, as is her habit, had been going strong since four o'clock that morning. Wojcicki was less sprightly, having just returned the previous night from her three-week honeymoon with Brin on safari in Africa and sailing around Greece and Turkey; she was also coming down with a nasty cold. After some idle chat about the biology of sleep, the board watched a demonstration of the company's user interface. Soon, the discussion turned to the thorny question of how much 23andMe will have to teach its customers about genetics to enable them to understand its offerings. "If we can get them to understand LD, that'll be an accomplishment," Avey said, referring to "linkage disequilibrium," a fairly obscure term describing how some genetic variations occur more often than anticipated. No, said Daphne Koller, a Stanford computer scientist and 2004 MacArthur fellow. "This should be a black box. LD is just going to trip them up."

As it happens, because 23andMe is a Web-based company, it can do both, letting the genetics hobbyist geek out on the details while giving the novice a minimum of information. Still, the challenge here was palpable: Starting a personal genomics company isn't like starting a Flickr or a Facebook. There's nothing intuitive about navigating your genome; it requires not just a new vocabulary but also a new conception of personhood. Scrape below the skin and we're flesh and bone; scrape below that and we're code. There's a massive amount of information to comprehend and fears to allay before customers will feel comfortable with the day-to-day utility of the site. 23andMe's solution is to offer a deep menu of FAQs, along with some nifty animation that explains the basic principles of genetics.

But the start-up is also careful not to overwhelm customers with foreboding information. Take its approach to monogenic conditions like Huntington's disease. For one thing, the company makes it clear that it is not in the diagnostic business and therefore doesn't provide specific genetic tests for specific diseases. But even if 23andMe wanted to, the SNP technology doesn't allow it, since many of the 1,400 monogenic conditions are diagnosed using techniques other than SNP testing. The BRCA1 and BRCA2 mutations that carry a high risk for breast cancer, for instance, are not SNPs but more complex defects that show up only in a test that sequences the entire gene. Similarly, the test for Huntington's looks for repeats of a certain nucleotide sequence rather than single-letter variations. Given the rarity of such conditions, it would be cost prohibitive to include these tests in a $1,000 run.

In other circumstances, the science is evolving so fast that 23andMe must invent a methodology as it goes. Take the essential task of calculating a customer's genetic risk for a disease, which the company delivers under its Odds Calculator. For a condition like type 2 diabetes, at least eight different SNPs have been correlated to the disease. Research among people of European descent has found that each of those SNPs has a slightly different effect—a variation of rs4712523 can increase one's risk by 17 percent, while a variation at rs7903146 can decrease risk by 15 percent. To crunch these numbers and determine one person's risk factor, 23andMe has opted to multiply the risks together. But a competing school of thought argues for adding the risk from SNP to SNP. The two approaches can result in wildly different tallies. "A lot of this is unknown. It's totally experimental," Wojcicki told me a few weeks before the science board meeting. "No one has looked at all eight diabetes markers together. They've all been identified individually, but they don't know exactly how they work together. So we've tried to make that clear."

All the ambiguity is indeed clear. There's no lack of caveats and in-context explanations on the site counseling customers to be cautious. In fact, the board at times even urged the company to hedge less and embrace the technology's gee-whiz factor, including uncertainty, more decisively. George Church, the Harvard geneticist who pioneered the sequencing techniques behind the Human Genome Project, sketched out a scenario: When a new study reporting a genetic association with a disease shows up in the *New York Times,* people are going to log on to 23andMe that morning and check to see whether the genetic marker in question is in their results. "People are going to wonder if you've got them covered," Church said. "And the answer better be yes."

In fact, that answer depends on the DNA chip that 23andMe uses to scan customer genomes. The company outsources that work to Illumina, the chip's developer. In its lab, Illumina extracts DNA from saliva and disperses it across a three-by-one-inch silicon wafer studded with more than 550,000 nanoscopic protein dots. Each dot detects a different SNP; more than half a million dots, strategically distributed across the human genome, cover a meaningful swath of anybody's DNA.

But it's possible that new research could turn up an association with a SNP that the 23andMe scan doesn't look for. And by definition, genotyping is a strategic, rather than an exhaustive, catalog.

The real endgame, therefore, is whole-genome sequencing, where you don't have to hope that you're covered—you'll know it. With whole-genome sequencing, all 3 billion base pairs of DNA will be identified: a complete library of your genetic code. As with DNA chips, sequencing technology is getting faster and costs are dropping. The Human Genome Project spent nearly $3 billion to sequence the first human genome. Sequencing DNA codiscoverer James Watson's genome cost just under $1 million; Craig Venter, who has already sequenced his genome at least once, is now spending about $300,000 to have it read again. Prices are expected to fall even more rapidly now that the X Prize Foundation has offered a $10 million award to the first team to sequence 100 human genomes in 10 days for less than $10,000 each.

At the board meeting, as talk turned to whole-genome sequencing, the energy in the room picked up. "This is absolutely the future," said Michael Eisen, a computational biologist at UC Berkeley. "It's exactly what the company should be doing as soon as possible."

"We will," responded Wojcicki, who then offered a juicy detail to the board. "We already have 10 people lined up and willing to pay $250,000 each for their whole genome. It's definitely something we want to do, maybe even in '08."

"George, how much will $250,000 get you?" Eisen asked Church, who's also on the X Prize advisory board. "How good a sequence would that be?"

"As good as Watson's," Church said. "At least as good."

Pushing the science forward is also a key part of the 23andMe business plan. As the company builds up its roster of customer genotypes, and later whole sequences, it gains a treasure trove of data that in turn can drive further research. On signing up, customers agree that their data, though still confidential, may be made available for scientific purposes. As the pool of participants grows, the start-up hopes to forge partnerships with academics and advocacy groups that focus on specific conditions. Already, the Parkinson's Institute is working with 23andMe on a study of Parkinson's disease. Similarly, 23andMe is talking with Autism Speaks, an advocacy group, about initiating research into autism—a disorder so complex that it will require the genetic information of many thousand research subjects to tease out potential associations.

This is also where a novel use of social-networking tools comes in. Wojcicki envisions groups of customers coming together around shared genotypes and SNPs, comparing notes about their conditions or backgrounds and identifying areas for further scientific research on their own. "It's a great way for individuals to be involved in the research world," Wojcicki says. "You'll have a profile, and something almost like a ribbon marking participation in these different research papers. It'll be like, 'How many *Nature* articles have you been part of?'" (Social networking will be included in version 2.0 in a matter of months, Avey says.)

For the board, such enterprising approaches to research are part of the fun of 23andMe. But after a long afternoon in a stuffy conference room, even geneticists can tire of too much genetics, and the meeting wound down. As the group walked into the foyer, someone asked about the two Segways there. Soon enough, some of the world's most celebrated geneticists had hopped aboard and were taking turns racing around the office at top speed.

My risk for heart disease may be lower than average, but that doesn't mean my genome isn't primed for problems. Far from it. Variations of three SNPs double my risk for prostate cancer, leaving me with a 30 percent chance of developing it in my lifetime. Restless legs syndrome, a dubious-sounding ailment characterized by jerky twitches in the middle of the night, was recently associated with a particular SNP variation—and I've got it, raising my risk by 32 percent. And my risk for exfoliating glaucoma, a type of eye disease, is a whopping three times the average American's. While the average person has just a 4 percent risk, my risk factor of 12 percent means it's something to mind.

Scanning my spreadsheet, all the odds start looking more like land mines. An 18 percent risk festers for this potentially fatal condition, a 13 percent risk ticks for that debilitating condition, and somewhere out there looms a 43 percent chance for something I may survive but sure don't want. And suddenly I realize: I can try to improve my odds here and there—eat less steak, schedule that colonoscopy earlier than most—but I'm going to go somehow, sometime. I can game the numbers, but I can't deny them.

Think of it this way: Health is an equation, with certain inputs and outputs. With conventional medicine, that means some fairly basic algebra: the simple addition and subtraction of symptoms and causes, with treatments like pharmaceuticals and surgery on the other end of the formula. For most Americans, the calculation results in fairly good health, with a life span stretching into the 70s. With the advent of genomics, though, we have stumbled into a far more arduous calculus, one requiring a full arsenal of algorithms and vectors. It's a more powerful tool—but it's also a lot more complicated.

It's not just the matter of accounting for all of our genetic markers and computing the attendant risk. That's just the start of it. Real personalized medicine must take into account traditional environmental factors, like smoking and diet and exercise. It also must consider the legion of pathogens out there, each with its own genetic quirks—not only the conventional ones of infectious disease but also the emerging class of viruses that seem to influence conditions from certain cancers to ulcers to obesity. Then there is the microbiome, the trillion-cell ecosystem of microbes that lives inside all of us, contributing to our health in largely mysterious ways. Oh, and save a piece of the equation for epigenetics, changes to the ways genes function without changes in the actual gene sequence. They contribute to our risk for common diseases such as cancer, heart disease, and diabetes.

Finally, leave a big blank spot for chance. No matter how much we learn from our genome, no matter how much it explains about us, randomness is always a looming factor in any health equation. Consider one behavior that is strongly associated with bad health—smoking. Everyone knows smoking is the single worst choice most people can make for their health. Yet the truth is that about a quarter of long-term smokers will not die of a smoking-related disease. Fate doesn't always work in our favor, though: Account for every known risk factor for heart disease—from high cholesterol to smoking to high blood pressure—and that explains only half the cases of the disease in the United States. In other words, I can bank on my genes and live in the most optimal way . . . and still die of a heart attack.

Mathematics isn't just a metaphor here. All of these variables are being broken down into data by scientists, and each data set is being scrutinized in an effort to quantify its impact on health. So let's make the leap of faith. The science is there, the data has been crunched, and it's all clear: Your genome is telling you that you face an elevated risk for certain diseases. What do you do? First, you likely go to your doctor (and let's assume she is one of the mere 800 MDs nationwide who has some training in genetics, so that she can actually make sense of your information). She considers your elevated risk and recommends some specific changes to your lifestyle. Will that work?

It might, if you act on that advice. But odds are you won't. In 1981, the National Institutes of Health completed a 10-year study that stands as the largest effort in scientific history to track behavior change. Starting with a pool of more than 360,000 Americans, the NIH set up centers around the country to study how well people would follow behaviors to alleviate the risk of heart disease. The subjects received personal counseling and support to help them stop smoking, eat better, and lose weight. At the study's end, though, 65 percent of the smokers still had the habit, half of those with high blood pressure still had it, and few had changed their diet at all. Subsequent studies have shown the same thing: Changing behavior is hard.

Luckily, there will be drugs tailored to work more effectively with our genetic quirks. These pharmacogenomics already exist: Herceptin specifically targets breast cancers that are caused by a growth protein from the HER2 gene, for instance, and more are in development. But taking a drug for several years, even one tailored to your DNA, can create a new set of disease risks and initiates a new trajectory of calculations.

The question becomes, then, whether you want to embark on this path of oddsmaking in the first place. Many individuals won't want to know what their genome has in store. Others will, only to join the worried well—those who live in fear of fulfilling their genetic destiny. And, of course, those genotyped or sequenced at birth won't have that choice; it'll already have been made for them.

Still, Wojcicki is onto something when she describes our genome as simply information. Already, we calibrate our health status in any number of ways, every day. We go to the drugstore and buy an HIV test or a pregnancy test. We take our blood pressure, track our cholesterol, count our calories. Our genome is now just one more metric at our disposal. It is one more factor revealed, an instrument suddenly within reach that can help us examine, and perhaps improve, our lives.

Walter Kirn

The Autumn of the Multitaskers

Neuroscience is confirming what we all suspect:
Multitasking is dumbing us down and driving us
crazy. One man's odyssey through the nightmare of
infinite connectivity.

> I think your suggestion is, Can we do two things
> at once? Well, we're of the view that we can
> walk and chew gum at the same time.
>   —*Richard Armitage,* deputy secretary of state,
>   on the wars in Afghanistan and Iraq, June 2,
>   2004 (Armitage announced his resignation on
>   November 16, 2004.)

> To do two things at once is to do neither.
>   —*Publilius Syrus,* Roman slave, first century BC

In the midwestern town where I grew up (a town so small
that the phone line on our block was a "party line" well into
the 1960s, meaning that we shared it with our neighbors and
couldn't use it while one of them was using it, unless we
wanted to quietly listen in—with their permission, naturally, and only if we were feeling awfully lonesome—while

they chatted with someone else), there were two skinny brothers in their 30s who built a car that could drive into the river and become a fishing boat.

My pals and I thought the car-boat was a wonder. A thing that did one thing but also did another thing—especially the *opposite* thing, but at least an *unrelated* thing—was our idea of a great invention and a bold stride toward the future. Where we got this idea, I'll never know, but it caused us to envision a world to come teeming with crossbred, hyphenated machines. Refrigerator-TV sets. Dishwasher-air conditioners. Table saw-popcorn poppers. Camera-radios.

With that last dumb idea, we were getting close to something, as I've noted every time I've dropped or fumbled my cell phone and snapped a picture of a wall or the middle button of my shirt. Impressive. Ingenious. Yet juvenile. Arbitrary. And why a substandard camera, anyway? Why not an excellent electric razor?

Because (I told myself at the cell phone store in the winter of 2003, as I handled a feature-laden upgrade that my new contract entitled me to purchase at a deep discount that also included a rebate) there may come a moment on a plane or in a subway station or at a mall when I and the other able-bodied males will be forced to subdue a terrorist, and my color snapshot of his trussed-up body will make the front page of *USA Today* and appear at the left shoulder of all the superstars of cable news.

While I waited for my date with citizen-journalist destiny, I took a lot of self-portraits in my Toyota and forwarded them to a girlfriend in Colorado, who reciprocated from her Jeep. Neither one of us almost died. For months. But then, one night on a snowy two-lane highway, while I was crossing Wyoming to see my girl's real face, my phone made its chirpy you-have-a-picture noise, and I glanced down in its direction while also, apparently, swerving off the pavement and sailing over a steep embankment toward a barbed-wire fence.

It was interesting to me—in retrospect, after having done some reading about the frenzied activity of the multi-tasking brain—how late in the process my prefrontal cortex, where our cognitive switchboards hide, changed its focus from the silly phone (*Where did it go? Did it slip between the seats? I wonder if this new photo is a nude shot or if it's another one from the topless series that seemed like such a breakthrough a month ago but now I'm getting sick of*) to the important matter of a steel fence post sliding spearlike across my hood . . .

(*But her arms are too short to shoot a nude self-portrait with a camera phone. She'd have to do it in a mirror . . .*)

The laminated windshield glass must have been high quality; the point of the post bounced off it, leaving only a star-shaped surface crack. But I was still barreling toward sagebrush, and who knew what rocks and boulders lay in wait . . .

Then the phone trilled out its normal ringtone.

Five minutes later, I'd driven out of the field and gunned it back up the embankment onto the highway and was proceeding south, heart slowing some, satellite radio tuned to a soft-rock channel called the Heart, which was playing lots of soothing Céline Dion.

"I just had an accident trying to see your picture."

"Will you get here in time to take me out to dinner?"

"I almost died."

"Well, you *sound* fine."

"Fine's not a *sound*."

I never forgave her for that detachment. I never forgave myself for buying a camera phone.

The abiding, distinctive feature of all crashes, whether in stock prices, housing values, or hit TV show ratings, is that they startle but don't surprise. When the euphoria subsides, when the volatile graph lines of excitability flatten and then curve down, people realize, collectively and instantly (and not infrequently with some relief), that they've been expecting this correction. The signs were everywhere, the warnings clear, the researchers in rough agreement, and the stories down at the bar and in the office (our own stories included) revealed the same anxieties.

Which explains why the busts and reversals we deem inevitable are also the least preventable and why they startle us, if briefly, when they come—because they were inevitable for so long that they should have come already. That they haven't, we reason, can mean only one of two things. Thanks to technology or some other magic, we've entered a new age when the laws of cause and effect (as propounded by Isaac Newton and Adam Smith) have yielded to the principle of dream-and-make-it-happen (as manifested by Steve Jobs and Oprah). Either that, or the thing that went up and up and up and hasn't come down, though it should have long ago, is being held aloft by our decision to forget it's up there and to carry on as though it weren't.

But on to the next inevitable contraction that everybody knows is coming, believes should have come a couple of years ago, and suspects can be postponed only if we pay no attention to the matter and stay very, very busy. I mean the end of the decade we may call the Roaring Zeros—these years of overleveraged, overextended, technology-driven, and finally unsustainable investment of our limited human energies in the dream of infinite connectivity. The overdoses, freak-outs, and collapses that converged in the late '60s to wipe out the gains of the wide-eyed optimists who set out to "Be Here Now" but ended up making posters that read "Speed Kills" are finally coming for the wired utopians who strove to "Be Everywhere at Once" but lost a measure of innocence, or should have, when their manic credo convinced us we could fight two wars at the same time.

The Multitasking Crash.

The Attention-Deficit Recession.

We all remember the promises. The slogans. They were all about freedom, liberation. Supposedly we were in handcuffs and wanted out of them. The key that dangled in front of us was a microchip.

"Where do you want to go today?" asked Microsoft in a mid-1990s ad campaign. The suggestion was that there were endless destinations—some geographic, some social, some intellectual—that you could reach in milliseconds by loading the right devices with the right software. It was further insinuated that where you went was purely up to you, not your spouse, your boss, your kids, or your government. Autonomy through automation.

This was the embryonic fallacy that grew up into the monster of multitasking.

Human freedom, as classically defined (to think and act and choose with minimal interference by outside powers), was not a product that firms like Microsoft could offer, but they recast it as something they *could* provide. A product for which they could raise the demand by refining its features, upping its speed, restyling its appearance, and linking it up with all the other products that promised freedom, too, but had replaced it with three inferior substitutes that they could market in its name:

Efficiency, convenience, and mobility.

For proof that these bundled minor virtues don't amount to freedom but are, instead, a formula for a period of mounting frenzy climaxing with a lapse into fatigue, consider that "Where do you want to go today?" was really manipulative advice, not an open question. "Go somewhere now," it strongly recommended, then go somewhere else to-morrow, but always go, go, go—and with our help. But did any rebel reply, "Nowhere. I like it fine right here"? Did anyone boldly ask, "What business is it of yours?" Was any-one brave enough to say, "Frankly, I want to go back to bed"?

Maybe a few of us. Not enough of us. Everyone else was going places, it seemed, and either we started going places, too—especially to those places that weren't *places* (another word they'd redefined) but were just pictures or documents or videos or boxes on screens where strangers conversed by typing—or else we'd be nowhere (a location once known as "here") doing nothing (an activity formerly labeled "liv-ing"). What a waste this would be. What a waste of our new freedom.

Our freedom to stay busy at all hours, at the task—and then the many tasks and ultimately the multitask—of trying to be free.

> While the president continued talking on the phone (Ms. Lewinsky understood that the caller was a Member of Congress or a Senator), she performed oral sex on him.
> —*The Starr Report,* 1998

It isn't working, it never has worked, and though we're still pushing and driving to make it work and puzzled as to why we haven't stopped yet, which makes us think we may go on forever, the stoppage or slowdown is coming nonethe-less, and when it does, we'll be startled for a moment, and then we'll acknowledge that, way down deep inside our-selves (a place that we almost forgot even existed), we always knew it *couldn't* work.

The scientists know this, too, and they think they know

why. Through a variety of experiments, many using functional magnetic resonance imaging to measure brain activity, they've torn the mask off multitasking and revealed its true face, which is blank and pale and drawn.

Multitasking messes with the brain in several ways. At the most basic level, the mental balancing acts that it requires—the constant switching and pivoting—energize regions of the brain that specialize in visual processing and physical coordination and simultaneously appear to short-change some of the higher areas related to memory and learning. We concentrate on the act of concentration at the expense of whatever it is that we're supposed to be concentrating *on*.

What does this mean in practice? Consider a recent experiment at UCLA, where researchers asked a group of 20-somethings to sort index cards in two trials, once in silence and once while simultaneously listening for specific tones in a series of randomly presented sounds. The subjects' brains coped with the additional task by shifting responsibility from the hippocampus—which stores and recalls information—to the striatum, which takes care of rote, repetitive activities. Thanks to this switch, the subjects managed to sort the cards just as well with the musical distraction—but they had a much harder time remembering what, exactly, they'd been sorting once the experiment was over.

Even worse, certain studies find that multitasking boosts the level of stress-related hormones such as cortisol and adrenaline and wears down our systems through biochemical friction, prematurely aging us. In the short term, the confusion, fatigue, and chaos merely hamper our ability to focus and analyze, but in the long term, they may cause it to atrophy.

The next generation, presumably, is the hardest hit. They're the ones way out there on the cutting edge of the multitasking revolution, texting and instant messaging each other while they download music to their iPod and update their Facebook page and complete a homework assignment and keep an eye on the episode of *The Hills* flickering on a nearby television. (A recent study from the Kaiser Family Foundation found that 53 percent of students in grades 7 through 12 report consuming some other form of media while watching television; 58 percent multitask while reading; 62 percent while using the computer; and 63 percent while listening to music. "I get bored if it's not all going at once," said a 17-year-old quoted in the study.) They're the ones whose still-maturing brains are being shaped to process information rather than understand or even remember it.

This is the great irony of multitasking—that its overall goal, getting more done in less time, turns out to be chimerical. In reality, multitasking slows our thinking. It forces us to chop competing tasks into pieces, set them in different piles, then hunt for the pile we're interested in, pick up its pieces, review the rules for putting the pieces back together, and then attempt to do so, often quite awkwardly. (Fact, and one more reason the bubble will pop: A brain attempting to perform two tasks simultaneously will, because of all the back-and-forth stress, exhibit a substantial lag in information processing.)

Productive? Efficient? More like running up and down a beach repairing a row of sand castles as the tide comes rolling in and the rain comes pouring down. *Multitasking,* a definition: "The attempt by human beings to operate like computers, often done with the assistance of computers." It begins by giving us more tasks to do, making each task harder to do, and dimming the mental powers required to do them. It finishes by making us forget exactly how on earth we did them (assuming we didn't give up, or "multiquit"), which makes them harder to do again.

Much of the problem is the metaphor. Or perhaps it's our need for metaphors in general, particularly when the subject is our minds and the comparison seems based on science. In the days of rudimentary chemistry, the mind was thought to be a beaker of swirling volatile essences. Then came classical physical mechanics, and the mind was regarded as a clock-like thing, with springs and wheels. Then it was steam driven, maybe. A combustion chamber. Then came electricity and Freud, and it was a dynamo of polarized energies—the id charged one way, the superego the other.

Now, in the heyday of the microchip, the brain is a computer. A CPU.

Except that it's not a CPU. It's whatever that thing is that's driven to misconstrue itself—over and over, century after century—as a prototype, rendered in all-too-vulnerable tissue, of our latest marvel of technology. And before the age of modern technology, *theology*. Further back than that, it's hard to voyage, since there was a period, common sense suggests, when we didn't even know we *had* brains. Or minds. Or spirits. Humans just sort of *did* stuff. And what they did was not influenced by metaphors about what they *ought* to be *capable* of doing but very well might not be equipped for (assuming you wanted to do it in the first place), like editing a playlist to e-mail to the lover whose husband you're interviewing on the phone about the movie he made that you're discussing in the blog entry you're posting tomorrow morning and are one-quarter watching certain parts of as you eat salad and carry on the call.

Would it be possible someday—through drugs, maybe, or esoteric Buddhism, or some profound, postapocalyptic languor—to stop coming up with ideas of what we are and then laboring to live up to them?

The great spooky splendor of the brain, of course, is that

no matter what we think it fundamentally resembles—even a small ethereal colosseum where angels smite demons and demons play dead, then suddenly spit fire into the angels' faces—it does a good job, a *great* job, of seeming to resemble it.

For a while.

I do like to read a book while having sex. And talk on the phone. You can get so much done.
　　—*Jennifer Connelly,* movie star, 2005

After the near-fatal consequences of my 2003 decision to buy a phone with a feature I didn't need, life went on, and rather rapidly, since multitasking eats up time in the name of saving time, rushing you through your two-year contract cycle and returning you to the company store with a suspicion that you didn't accomplish all you hoped to after your last optimistic, euphoric visit.

"Which of the ones that offer rebates don't have cameras in them?"

"The decent models all do. The best ones now have video capabilities. You can shoot little movies."

I wanted to ask, *Of what? Oncoming barbed wire?* The salesman was a believer, though—a zealot.

"Oh, yeah," he said, "as well as GPS-based, turn-by-turn navigation systems. Which are cool if you drive a lot."

"You have to look down at the screen, though."

"They're paid subscription services, you need to know, but we're giving away the first month free, and even after that, the rates are reasonable."

I shook my head. I was turning down whiz-bang features for the first time, and so had some of my friends, one of whom had sprung for a new BlackBerry that he'd holed up in his office to learn to use. He'd emerged a week later looking demoralized, muttering about getting old, although he'd just turned 34.

"Those little ones there—the ones that aren't so slim, that you give away free."

"That too is an option. Mostly they're aimed at kids, though. Adolescents."

I wanted one anyway. I'd caught air in my Land Cruiser off a sheer embankment, lost my girlfriend, chucked my dream of snapping a hog-tied terrorist, and once, because of another girl—a jealous type who never trusted that I was where I said I was—been forced to send on a shot of L.A. palm trees to prove that I was not in Oregon meeting up with yet another girl whom I'd drunk coffee with after a poetry reading and who must have been bombed a few weeks later when she sent me a text message at 3:00 a.m. while I was sleeping beside the jealous girl. My bedmate heard the ring, crept out of bed, and read the message, then woke me up and demanded that I explain why it seemed to suggest we'd shared more than double espressos—an effect curiously enhanced by the note's thumb-typed dyslexic style: *Thuoght I saw thoes parkly eyes this aft, that sensaul deivlish mouth, and it took me rihgt in again, like vapmires do.*

"I'll take the fat little free one," I told the salesman.

"The thing's inert. It does nothing. It's a pet rock."

I informed him that I was old enough to have actually owned a pet rock once and that I missed it.

Here's the worst of the chilling little thoughts that have come to me during micro-tasking seize-ups: For every driver who's ever died while talking on a cell phone (researchers at the Harvard Center for Risk Analysis estimate that some 2,600 deaths and 330,000 injuries may be caused by drivers on cell phones each year), there was someone on the other end who, chances are, was too distracted to notice. Too busy cooking, NordicTracking, fluffing up his online dating profile, or—most hauntingly of all, I'd think, for a listener destined to discover that the acoustic chaos he'd interpreted as the other phone going out of range, or perhaps as a networkwide disturbance triggered by a solar flare, was actually a death, a human death, a death he had some role in—sitting on the toilet.

Trading securities.

Or would watching streaming pornography be worse?

Not that both of these activities can't be performed on the same computer screen. And often are—you can bet on it. In bathrooms. Even *airport* bathrooms, on occasion. In some of which, via radio, the latest business headlines can be monitored, permitting (in theory and therefore in *fact,* because, as the First Law of Multitasking dictates, any 2 or 8 or 16 processes that *can* overlap *must* overlap) the squatting day trader viewing the dirty webcast (while on the phone with someone, don't forget) to learn that the company he just bought stock in has entered merger talks with *his own employer* and surged almost 20 percent in under three minutes!

"Guess how much richer I've gotten while we've been yakking?" he says into his cell, breaking his own rule about pretending that when he's on the phone, he's on the phone. Exclusively. Fully. With his entire being.

No reply.

Must be driving through a tunnel.

I've been fired, I've been insulted in front of coworkers, but the time I flew thousands of miles to meet a boss who spent our first and only hour together politely nodding at my proposals while thumbing out messages on a new device, whose existence neither of us acknowledged and whose screen he kept tilted so I couldn't see it, still feels, five years later, like the low point of my career.

> This is the perfect "one plus one equals three" opportunity.
>> —*Robert Pittman,* president and COO of America Online, on the merger between AOL and Time Warner, 2000

There may be a financial cost to multitasking as well. The sum is extremely large and hard to vouch for, the esoteric algorithm that yielded it a puzzle to all but its creator, possibly, but it's one of those figures that's fun to quote in bars.

Six hundred and fifty billion dollars. That's what we might call our National Attention Deficit, according to Jonathan B. Spira, who's the chief analyst at a business-research firm called Basex and who has estimated the per annum cost to the economy of multitasking-induced disruptions. (He obtained the figure by surveying office workers across the country, who reported that some 28 percent of their time was wasted dealing with multitasking-related transitions and interruptions.)

That $650 billion reflects just one year's loss. This means that the total debt is vastly higher, since personal digital assistants (the devices that, in my opinion, turned multitasking from a habit into a pathology, which the advent of Bluetooth then rendered fatal and the spread of wireless broadband made communicable) are several annums old. This puts our shortfall somewhere in the trillions—even before we add in the many billions that vanished when Time Warner and AOL joined their respective corporate missions—so ably accomplished when the firms were separate—into one colossal mission impossible.

And don't forget to add Enron to the tab, a company that seemed to master so many enterprises, trading everything from energy to weather futures, that the Wall Street analysts' brains froze up trying to "recontext" (another science term) what looked at first like a capitalist dynamo as the street-corner con that it turned out to be. Reports suggest that the illusion depended nearly as much on cunning set design as it did on phony accounting. The towering stack of Broadway stages that Enron called its headquarters—with its profusion of workstations, trading boards, copiers, speakerphones, fax machines, and shredders—made visiting banker-broker types go snow-blind. When the fraud was exposed, the press accused the moneymen of overestimating Enron. In truth, they'd underestimated Enron, whose hectic multitasking front concealed the managers' Zenlike focus on one proficiency, and only one.

Hypnotism.

Which is easy to practice on an audience whose brains are already half dormant from the stress of scheduling flights on fractionally owned jets and changing the tilt and speed of treadmills according to the shifting readouts of miniature biofeedback monitors strapped around their upper arms.

What has the madness of multitasking cost us? The better question might be: What hasn't it?

And the IOUs keep coming, signed at the bottom with millions of our names. We issued this currency. We're the Federal Reserve of the attention economy, the central bank of overcommitment, keeping the system liquid with adrenaline. The problem is that we, the bankers, are also the borrowers. That's multitasking for you. It moves in circles.

Circles that we run around ourselves, as we try to pay off the debts we owe ourselves with funny money engraved with our own faces.

Here's one item from my ledger:

Cost of pitying Kevin Federline while organizing business trip online and attaching computer peripheral: $279.

Federline—I know. A mayfly on the multimedia river who, now that he has mated, deserves to break back up into pixels. That he hasn't means pixels are far too cheap and plentiful, particularly on the AOL welcome page, where for several months last year Federline's image was regularly positioned beside the icon I click to get my e-mail. With practice, I learned to sweep past him the way the queen sweeps past her guards, but one afternoon his picture triggered a brainslide that buried half my day.

What the avalanche overwhelmed was a mental function that David E. Meyer, a psychology professor at the University of Michigan, calls "adaptive executive control." Thanks to Federline, I lost my ability, as Meyer would say, to "schedule task processes appropriately" and to "obey instructions about their relative priorities."

Meyer, it's worth noting, is a relative optimist among the researchers studying multitasking, since he's convinced that some people can learn, with enough practice, to perform two tasks simultaneously as successfully as if they were doing them sequentially. But "enough practice" turns out to mean at least 2,000 tries, and I had just the one chance at the cheap fare to San Francisco that I'd turned on my laptop to reserve, only to be distracted by the picture of Federline winking at me from one browser window over.

The photo, a link explained, was taken while Federline was taping a TV show and happened to peer down at his phone, only to learn that what's-her-hair, his wife, the psycho, bad-mother rehab-escapee (I had last caught up on her misadventures weeks or months before, while waiting out an eBay auction for an auxiliary hard drive "still in box"), had sent him a text message asking for a divorce. Federline's face looked as raw as a freshly unbandaged plastic-surgery patient's, but the aspect of the photo that grabbed me (as the promotional fare hovered in the ether, still unbooked and up for grabs) was the idea I suddenly entertained about its origins. The picture of Federline in cell phone shock had been snapped on the sly by another phone, I sensed, and possibly by a hanger-on whom Federline regarded as a "bro." It also seemed plausible that after the taping, Federline bought dinner for this Judas—who, in my reconstruction of events, had already beamed the spy shot to a tabloid and been wired big money in return. If so, he was probably richer than Federline, who depended for funds on the wife who'd just dumped him.

This thought sequence caused me to remember the hard drive—still sitting unopened in a closet—that I'd bought in that internet auction way back when, while catching up on the Hollywood gossip news. Here's the mental flowchart: Federline dumped > story about his prenuptial with Britney Spears > story was read during eBay auction > time to get some use out of my purchase.

Removing the hard drive from its shell of molded Styrofoam sloppily wrapped in masking tape stirred serious doubts about the seller's claim that the gadget was unused. This put me in a quandary. Should I send the hard drive back? Blackball its seller on a message board? Best to test it first. I riffled through drawers to find the proper cable, plugged the device into a USB port, and only then became aware of the fluorescent Post-it note stuck in the corner of my laptop screen. "Grab discount SF fare," the note read. Where had it gone? Where had *I* gone, rather? How could a piece of paper in a color specially formulated to signal the brain *Important! Don't Ignore!* be upstaged by a picture of a sad minor celebrity? If the Post-it note had been a road sign warning of a hairpin mountain curve and Federline's photo a radio interview, I and my car would be rolling down a cliff now.

Back to the San Francisco ticket, then. I brought up the main Expedia/Orbitz/Travelocity page and typed in the code for the San Francisco airport, which I couldn't believe I got wrong. To fix it, I was forced to use one of those drop-down alphabetized lists that the highlight line always moves too fast through, meaning I click my mouse several entries too late. Seattle this time. I scrolled back up.

All tickets sold out.

The scientists call this ruinous mental lurching "dual task interference," or just plain bottlenecking. I call it the reason Keven Federline cost me a cheap flight to San Francisco. (It also explains, perhaps, why sexual threesomes are often disappointing.)

I just wish the military understood the concept. They might understand then why "walking and chewing gum" in Afghanistan and Iraq is no way to catch bin Laden.

My hunch is that when we look back on it someday, at our juggling of electronic lives and the array of subtly different personas that each one encourages (we're terse when texting, freewheeling on the phone, and in some middle state while e-mailing), the spectacle will appear as quaint and stylized as those scenes in old movies of stiff-backed lady operators, hair in bobby pins, rapidly swapping phone jacks from hole to hole as they connect Chicago to Miami, reporter to city desk, businessman to mistress. Such scenes were, for a time, cinematic shorthand for the frenzy of modern life, but then communications technology changed, and those operators lost their jobs.

To us.

We've got to be patient and committed [in Iraq], but we've got to multitask. . . . We've got to talk about Iran—Iran is more dangerous than Iraq—*and we have got to get the job done in Afghanistan and in Pakistan.*

—*Rudolph Giuliani,* Republican presidential candidate, July 2007

The night the bubble finally popped for me began when I pushed a button on my hospital bed to summon the gray-haired night nurse. To convey my appreciation when she arrived and to help establish a relationship that I hoped would lead her to agree with me that my morphine drip was far too slow, I did as the gurus of management urge executives to do when they engage in important negotiations. I "reallocated" my "presence" and "enriched" my "medium." I removed my headphones, closed my book, aimed the remote and clicked off the TV, and looked the old woman in the eye.

"What?" she said.

Her question came too quickly. Because of the way the human brain works—always lagging slightly, always falling a bit behind itself when it has to drop many things, one thing at a time, and refocus on a new thing—my attention had not yet caught up with my expression. Also, perhaps because of the way that morphine works, I was unnaturally aware of the mechanisms inside my mind. I could actually feel the neurological switching, the mental grinding of fine, tiny gears that makes multitasking such an inefficient, slow, error-prone, tiring way to get things done.

"Still hurts," I finally said. "Wondering if you'd shorten up the intervals." I left out the *I*'s, text message–style, because that's how people in agony communicate. Teenagers, too, but aren't they also in agony, with the shy self-consciousness of partials who don't show all their cards, out of fear that they haven't yet drawn many worth playing?

The nurse made a face that the gurus would call "equivocal"—meaning that it can support conflicting interpretations, even in a real-time, face-to-face, "presence-rich" exchange—and then glanced down at the iPod on my blanket.

"Music lover?"

"Book on tape," I said.

"You can do those both at once?" She eyed the real book lying on my lap.

"Same one," I said. "I like to double up."

"Why?"

I had no answer. I had a comeback—*Because I can, because it's possible*—but a comeback is just a way to keep things rolling when perhaps they ought to stop. When the nurse looked away and punched in new instructions on the keypad attached to my IV stand, I heard her thinking, *No wonder this guy has kidney stones. No wonder he's so hungry for narcotics.* She turned around in time to see my hands moving from the book they'd just reopened to the tangled wires of the earphones.

"I'm grateful that you came so quickly and showed such understanding," I said, not textishly, relaxing my syntax to suit the expectations of the elderly.

"Maybe more dope will be just the thing," the nurse said, shedding equivocation with every word, as a dreamy warmth spread through my limbs and she soft-stepped out and shut the door. When I woke in the wee hours, my book, in both its forms, had slid off the bed onto the floor, the TV remote was lost among the blankets, and the blinking "sleep" indicator of the laptop computer I've failed to mention (delivered to my bedside by a friend who'd shared my delusion that even 25-bed Montana hospitals must offer wireless internet these days) was exhaling onto the walls a lovely blue light that tempted me never to boot it up again.

That night, last May, as I drowsed and passed my stones, the mania left me, and it hasn't returned.

What happened to the skinny brothers' car-boat was that it sank the third time they took it fishing. It cracked down the length of its hull, took on water, and then nose-dived into the sandy bottom, leaving its revved-up rear propeller sticking up two feet out of the river, furiously churning air until its creators returned in a canoe and whacked it silent with a crowbar.

The catastrophe, visible from half the town, was the talk of the party line that night, with most of the grown-ups joining in one pooled call that was still humming when I was sent to bed.

"Where do you want to go today?" Microsoft asked us.

Now that I no longer confuse freedom with speed, convenience, and mobility, my answer would be: "Away. Just away. Someplace where I can think."

Jeffrey Rosen

The Brain on the Stand

Neuroscience is becoming the newest witness for the defense.

I. MR. WEINSTEIN'S CYST

When historians of the future try to identify the moment that neuroscience began to transform the American legal system, they may point to a little-noticed case from the early 1990s. The case involved Herbert Weinstein, a 65-year-old ad executive who was charged with strangling his wife, Barbara, to death and then, in an effort to make the murder look like a suicide, throwing her body out the window of their 12th-floor apartment on East 72nd Street in Manhattan. Before the trial began, Weinstein's lawyer suggested that his client should not be held responsible for his actions because of a mental defect—namely, an abnormal cyst nestled in his arachnoid membrane, which surrounds the brain like a spider web.

The implications of the claim were considerable. American law holds people criminally responsible unless they act under duress (with a gun pointed at the head, for example) or they suffer from a serious defect in rationality—like not being able to tell right from wrong. But if you suffer from such a serious defect, the law generally doesn't care why—whether it's an unhappy childhood or an arachnoid cyst or both. To suggest that criminals could be excused because their brains made them do it seems to imply that anyone whose brain isn't functioning properly could be absolved of responsibility. But should judges and juries really be in the business of defining the normal or properly working brain? And since all behavior is caused by our brains, wouldn't this mean all behavior could potentially be excused?

The prosecution at first tried to argue that evidence of Weinstein's arachnoid cyst shouldn't be admitted in court. One of the government's witnesses, a forensic psychologist named Daniel Martell, testified that brain-scanning technologies were new and untested and their implications weren't yet widely accepted by the scientific community. Ultimately, on October 8, 1992, Judge Richard Carruthers issued a Solomonic ruling: Weinstein's lawyers could tell the jury that brain scans had identified an arachnoid cyst, but they couldn't tell jurors that arachnoid cysts were associated with violence. Even so, the prosecution team seemed to fear that simply exhibiting images of Weinstein's brain in court would sway the jury. Eleven days later, on the morning of jury selection, they agreed to let Weinstein plead guilty in exchange for a reduced charge of manslaughter.

After the Weinstein case, Daniel Martell found himself in so much demand to testify as an expert witness that he started a consulting business called Forensic Neuroscience. Hired by defense teams and prosecutors alike, he has testified over the past 15 years in several hundred criminal and civil cases. In those cases, neuroscientific evidence has been admitted to show everything from head trauma to the tendency of violent video games to make children behave aggressively. But Martell told me that it's in death penalty litigation that neuroscience evidence is having its most revolutionary effect. "Some sort of organic brain defense has become de rigueur in any sort of capital defense," he said. Lawyers routinely order scans of convicted defendants' brains and argue that a neurological impairment prevented them from controlling themselves. The prosecution counters that the evidence shouldn't be admitted, but under the relaxed standards for mitigating evidence during capital sentencing, it usually is. Indeed, a Florida court has held that the failure to admit neuroscience evidence during capital sentencing is grounds for a reversal. Martell remains skeptical about the worth of the brain scans, but he observes that they've "revolutionized the law."

The extent of that revolution is hotly debated, but the influence of what some call neurolaw is clearly growing. Neuroscientific evidence has persuaded jurors to sentence defendants to life imprisonment rather than to death; courts have also admitted brain-imaging evidence during criminal trials to support claims that defendants like John W. Hinckley Jr., who tried to assassinate President Reagan, are insane. Carter Snead, a law professor at Notre Dame, drafted a staff working paper on the impact of neuroscientific evidence in criminal law for President Bush's Council on Bioethics. The report concludes that neuroimaging evidence is of mixed reliability, but "the large number of cases in which such evidence is presented is striking." That number will no doubt increase substantially. Proponents of neurolaw say that neuroscientific evidence will have a large impact not only on questions of guilt and punishment but also on the detection of lies and hidden bias and on the prediction of future criminal behavior. At the same time, skeptics fear that the use of brain-scanning technology as a kind of super mind-reading device will threaten our privacy and mental freedom, leading some to call for the legal system to respond with a new concept of "cognitive liberty."

One of the most enthusiastic proponents of neurolaw is Owen Jones, a professor of law and biology at Vanderbilt. Jones (who happens to have been one of my law school classmates) has joined a group of prominent neuroscientists and law professors who have applied for a large MacArthur Foundation grant; they hope to study a wide range of neurolaw questions, such as Do sexual offenders and violent teenagers show unusual patterns of brain activity? or Is it possible to capture brain images of chronic neck pain when someone claims to have suffered whiplash? In the meantime, Jones is turning Vanderbilt into a kind of Los Alamos for neurolaw. The university has just opened a $27 million neuroimaging center and has poached leading neuroscientists from around the world; soon, Jones hopes to enroll students in the nation's first program in law and neuroscience. "It's breathlessly exciting," he says. "This is the new frontier in law and science—we're peering into the black box to see how the brain is actually working, that hidden place in the dark quiet, where we have our private thoughts and private reactions—and the law will inevitably have to decide how to deal with this new technology."

II. A VISIT TO VANDERBILT

Owen Jones is a disciplined and quietly intense man, and his enthusiasm for the transformative power of neuroscience is infectious. With René Marois, a neuroscientist in the psychology department, Jones has begun a study of how the human brain reacts when asked to impose various punishments. Informally, they call the experiment Harm and Punishment—and they offered to make me one of their first subjects.

We met in Jones's pristine office, which is decorated with a human skull and calipers, like those that phrenologists once used to measure the human head; his father is a dentist, and his grandfather was an electrical engineer who collected tools. We walked over to Vanderbilt's Institute of Imaging Science, which, although still surrounded by scaffolding, was as impressive as Jones had promised. The basement contains one of the few 7-tesla magnetic resonance imaging scanners in the world. For Harm and Punishment, Jones and Marois use a less powerful 3 tesla, which is the typical research MRI.

We then made our way to the scanner. After removing all metal objects—including a belt and a stray dry-cleaning tag with a staple—I put on earphones and a helmet that was shaped like a birdcage to hold my head in place. The lab assistant turned off the lights and left the room; I lay down on the gurney and, clutching a panic button, was inserted into the magnet. All was dark except for a screen flashing hypothetical crime scenarios, like this one: "John, who lives at home with his father, decides to kill him for the insurance money. After convincing his father to help with some electrical work in the attic, John arranges for him to be electrocuted. His father survives the electrocution, but he is hospitalized for three days with injuries caused by the electrical shock." I was told to press buttons indicating the appropriate level of punishment, from zero to nine, as the magnet recorded my brain activity.

After I spent 45 minutes trying not to move an eyebrow while assigning punishments to dozens of sordid imaginary criminals, Marois told me through the intercom to try another experiment: namely, to think of familiar faces and places in sequence, without telling him whether I was starting with faces or places. I thought of my living room, my wife, my parents' apartment, and my twin sons, trying all the while to avoid improper thoughts for fear they would be discovered. Then the experiments were over, and I stumbled out of the magnet.

The next morning, Owen Jones and I reported to René Marois's laboratory for the results. Marois's graduate students, who had been up late analyzing my brain, were smiling broadly. Because I had moved so little in the machine, they explained, my brain activity was easy to read. "Your head movement was incredibly low, and you were the harshest punisher we've had," Josh Buckholtz, one of the grad students, said with a happy laugh. "You were a researcher's dream come true!" Buckholtz tapped the keyboard, and a high-resolution 3-D image of my brain appeared on the screen in vivid colors. Tiny dots flickered back and forth, showing my eyes moving as they read the lurid criminal scenarios. Although I was only the fifth subject to be put in the scanner, Marois emphasized that my punishment ratings were higher than average. In one case, I assigned a seven where the average punishment was four. "You were focusing on the intent, and the others focused on the harm," Buckholtz said reassuringly.

Marois explained that he and Jones wanted to study the interactions among the emotion-generating regions of the brain, like the amygdala, and the prefrontal regions responsible for reason. "It is also possible that the prefrontal cortex is critical for attributing punishment, making the essential decision about what kind of punishment to assign," he suggested. Marois stressed that, in order to study that possibility, more subjects would have to be put into the magnet. But if the prefrontal cortex does turn out to be critical for selecting among punishments, Jones added, it could be highly relevant for lawyers selecting a jury. For example, he suggested, lawyers might even select jurors for different cases based on their different brain-activity patterns. In a complex insider-trading case, for example, perhaps the defense would "like to have a juror making decisions on maximum deliberation and minimum emotion"; in a government entrapment case, emotional reactions might be more appropriate.

We then turned to the results of the second experiment, in which I had been asked to alternate between thinking of faces and places without disclosing the order. "We think we can guess what you were thinking about, even though you didn't tell us the order you started with," Marois said proudly. "We think you started with places, and we will prove to you that it wasn't just luck." Marois showed me a picture of my parahippocampus, the area of the brain that responds strongly to places and the recognition of scenes. "It's lighting up like Christmas on all cylinders," Marois said. "It worked beautifully, even though we haven't tried this before here."

He then showed a picture of the fusiform area, which is responsible for facial recognition. It, too, lit up every time I thought of a face. "This is a potentially very serious legal implication," Jones broke in, since the technology allows us to tell what people are thinking about even if they deny it. He pointed to a series of practical applications. Because subconscious memories of faces and places may be more reliable than conscious memories, witness lineups could be transformed. A child who claimed to have been victimized by a stranger, moreover, could be shown pictures of the faces of suspects to see which one lit up the face-recognition area in ways suggesting familiarity.

Jones and Marois talked excitedly about the implications of their experiments for the legal system. If they discovered a significant gap between people's hard-wired sense of how severely certain crimes should be punished and the actual punishments assigned by law, federal sentencing guidelines might be revised, on the principle that the law shouldn't diverge too far from deeply shared beliefs. Experiments might help to develop a deeper understanding of the criminal brain or of the typical brain predisposed to criminal activity.

III. THE END OF RESPONSIBILITY?

Indeed, as the use of functional MRI results becomes increasingly common in courtrooms, judges and juries may be asked to draw new and sometimes troubling lines between "normal" and "abnormal" brains. Ruben Gur, a professor of psychology at the University of Pennsylvania School of Medicine, specializes in doing just that. Gur began his expert-witness career in the mid-1990s when a colleague asked him to help in the trial of a convicted serial killer in Florida named Bobby Joe Long. Known as the "classified-ad rapist," because he would respond to classified ads placed by women offering to sell household items and then rape and kill them, Long was sentenced to death after he committed at least nine murders in Tampa. Gur was called as a national expert in positron-emission tomography, or PET scans, in which patients are injected with a solution containing radioactive markers that illuminate their brain activity. After examining Long's PET scans, Gur testified that a motorcycle accident that had left Long in a coma had also severely damaged his amygdala. It was after emerging from the coma that Long committed his first rape.

"I didn't have the sense that my testimony had a profound impact," Gur told me recently—Long is still filing appeals—but he has testified at more than 20 capital cases since then. He wrote a widely circulated affidavit arguing that adolescents are not as capable of controlling their impulses as adults because the development of neurons in the prefrontal cortex isn't complete until the early 20s. Based on that affidavit, Gur was asked to contribute to the preparation of one of the briefs filed by neuroscientists and others in *Roper v. Simmons,* the landmark case in which a divided Supreme Court struck down the death penalty for offenders who committed crimes when they were under the age of 18. The leading neurolaw brief in the case, filed by the American Medical Association and other groups, argued that because "adolescent brains are not fully developed" in the prefrontal regions, adolescents are less able than adults to control their impulses and should not be held fully accountable "for the immaturity of their neural anatomy." In his majority decision, Justice Anthony Kennedy declared that "as any parent knows and as the scientific and sociological studies" cited in the briefs "tend to confirm, '[a] lack of maturity and an underdeveloped sense of responsibility are found in youth more often than in adults.'" Although Kennedy did not cite the neuroscience evidence specifically, his indirect reference to the scientific studies in the briefs led some supporters and critics to view the decision as the *Brown v. Board of Education* of neurolaw.

One important question raised by the *Roper* case was where to draw the line in considering neuroscience evidence as a legal mitigation or excuse. Should courts be in the business of deciding when to mitigate someone's criminal responsibility because his brain functions improperly, whether because of age, in-born defects, or trauma? As we learn more about criminals' brains, will we have to redefine our most basic ideas of justice?

Two of the most ardent supporters of the claim that neuroscience requires the redefinition of guilt and punishment are Joshua D. Greene, an assistant professor of psychology at Harvard, and Jonathan D. Cohen, a professor of psychology who directs the neuroscience program at Princeton. Greene got Cohen interested in the legal implications of neuroscience, and together they conducted a series of experiments exploring how people's brains react to moral dilemmas involving life and death. In particular, they wanted to test people's responses in the fMRI scanner to variations of the famous trolley problem, which philosophers have been arguing about for decades.

The trolley problem goes something like this: Imagine a train heading toward five people who are going to die if you don't do anything. If you hit a switch, the train veers onto a side track and kills another person. Most people confronted with this scenario say it's OK to hit the switch. By contrast, imagine that you're standing on a footbridge that spans the train tracks, and the only way you can save the five people is to push an obese man standing next to you off the footbridge so that his body stops the train. Under these circumstances, most people say it's not OK to kill one person to save five.

"I wondered why people have such clear intuitions," Greene told me, "and the core idea was to confront people with these two cases in the scanner and see if we got more of an emotional response in one case and reasoned response in the other." As it turns out, that's precisely what happened: Greene and Cohen found that the brain region associated with deliberate problem solving and self-control, the dorsolateral prefrontal cortex, was especially active when subjects confronted the first trolley hypothetical, in which most of them made a utilitarian judgment about how to save the greatest number of lives. By contrast, emotional centers in the brain were more active when subjects confronted the second trolley hypothetical, in which they tended to recoil at the idea of personally harming an individual, even under such wrenching circumstances. "This suggests that moral judgment is not a single thing; it's intuitive emotional responses and then cognitive responses that are duking it out," Greene said.

"To a neuroscientist, you are your brain; nothing causes

your behavior other than the operations of your brain," Greene says. "If that's right, it radically changes the way we think about the law. The official line in the law is all that matters is whether you're rational, but you can have someone who is totally rational but whose strings are being pulled by something beyond his control." In other words, even someone who has the illusion of making a free and rational choice between soup and salad may be deluding himself, since the choice of salad over soup is ultimately predestined by forces hard-wired in his brain. Greene insists that this insight means that the criminal justice system should abandon the idea of retribution—the idea that bad people should be punished because they have freely chosen to act immorally—which has been the focus of American criminal law since the 1970s, when rehabilitation went out of fashion. Instead, Greene says, the law should focus on deterring future harms. In some cases, he supposes, this might mean lighter punishments. "If it's really true that we don't get any prevention bang from our punishment buck when we punish that person, then it's not worth punishing that person," he says. (On the other hand, Carter Snead, the Notre Dame scholar, maintains that capital defendants who are not considered fully blameworthy under current rules could be executed more readily under a system that focused on preventing future harms.)

Others agree with Greene and Cohen that the legal system should be radically refocused on deterrence rather than on retribution. Since the celebrated M'Naughten case in 1843, involving a paranoid British assassin, English and American courts have recognized an insanity defense only for those who are unable to appreciate the difference between right and wrong. (This is consistent with the idea that only rational people can be held criminally responsible for their actions.) According to some neuroscientists, that rule makes no sense in light of recent brain-imaging studies. "You can have a horrendously damaged brain where someone knows the difference between right and wrong but nonetheless can't control their behavior," says Robert Sapolsky, a neurobiologist at Stanford. "At that point, you're dealing with a broken machine, and concepts like punishment and evil and sin become utterly irrelevant. Does that mean the person should be dumped back on the street? Absolutely not. You have a car with the brakes not working, and it shouldn't be allowed to be near anyone it can hurt."

Even as these debates continue, some skeptics contend that both the hopes and fears attached to neurolaw are overblown. "There's nothing new about the neuroscience ideas of responsibility; it's just another material, causal explanation of human behavior," says Stephen J. Morse, professor of law and psychiatry at the University of Pennsylvania. "How is this different than the Chicago school of sociology," which tried to explain human behavior in terms of environment and social structures? "How is it different from genetic explanations or psychological explanations? The only thing different about neuroscience is that we have prettier pictures and it appears more scientific."

Morse insists that "brains do not commit crimes; people commit crimes"—a conclusion he suggests has been ignored by advocates who, "infected and inflamed by stunning advances in our understanding of the brain . . . all too often make moral and legal claims that the new neuroscience . . . cannot sustain." He calls this "brain overclaim syndrome" and cites as an example the neuroscience briefs filed in the Supreme Court case *Roper v. Simmons* to question the juvenile death penalty. "What did the neuroscience add?" he asks. If adolescent brains caused all adolescent behavior, "we would expect the rates of homicide to be the same for 16- and 17-year-olds everywhere in the world—their brains are alike—but in fact, the homicide rates of Danish and Finnish youths are very different than American youths." Morse agrees that our brains bring about our behavior—"I'm a thoroughgoing materialist, who believes that all mental and behavioral activity is the causal product of physical events in the brain"—but he disagrees that the law should excuse certain kinds of criminal conduct as a result. "It's a total non sequitur," he says. "So what if there's biological causation? Causation can't be an excuse for someone who believes that responsibility is possible. Since all behavior is caused, this would mean all behavior has to be excused." Morse cites the case of Charles Whitman, a man who, in 1966, killed his wife and his mother and then climbed up a tower at the University of Texas and shot and killed 13 more people before being shot by police officers. Whitman was discovered after an autopsy to have a tumor that was putting pressure on his amygdala. "Even if his amygdala made him more angry and volatile, since when are anger and volatility excusing conditions?" Morse asks. "Some people are angry because they had bad mommies and daddies and others because their amygdalas are mucked up. The question is: When should anger be an excusing condition?"

Still, Morse concedes that there are circumstances under which new discoveries from neuroscience could challenge the legal system at its core. "Suppose neuroscience could reveal that reason actually plays no role in determining human behavior," he suggests tantalizingly. "Suppose I could show you that your intentions and your reasons for your actions are post hoc rationalizations that somehow your brain generates to explain to you what your brain has already done" without your conscious participation. If neuroscience could reveal us to be automatons in this respect, Morse is prepared to agree with Greene and Cohen that criminal law would have to abandon its current ideas about responsibility and seek other ways of protecting society.

Some scientists are already pushing in this direction. In a series of famous experiments in the 1970s and '80s, Benjamin Libet measured people's brain activity while telling them to move their fingers whenever they felt like it. Libet detected brain activity suggesting a readiness to move the finger half a second before the actual movement and about 400 milliseconds before people became aware of their conscious intention to move their finger. Libet argued that this leaves 100 milliseconds for the conscious self to veto the brain's unconscious decision or to give way to it—suggesting, in the words of the neuroscientist Vilayanur S. Ramachandran, that we have not free will but "free won't."

Morse is not convinced that the Libet experiments reveal us to be helpless automatons. But he does think that the study of our decision-making powers could bear some fruit for the law. "I'm interested," he says, "in people who suffer from drug addictions, psychopaths and people who have intermittent explosive disorder—that's people who have no general rationality problem other than they just go off." In other words, Morse wants to identify the neural triggers that make people go postal. "Suppose we could show that the higher deliberative centers in the brain seem to be disabled in these cases," he says. "If these are people who cannot control episodes of gross irrationality, we've learned something that might be relevant to the legal ascription of responsibility." That doesn't mean they would be let off the hook, he emphasizes: "You could give people a prison sentence and an opportunity to get fixed."

IV. PUTTING THE UNCONSCIOUS ON TRIAL

If debates over criminal responsibility long predate the fMRI, so do debates over the use of lie-detection technology. What's new is the prospect that lie detectors in the courtroom will become much more accurate and, correspondingly, more intrusive. There are, at the moment, two lie-detection technologies that rely on neuroimaging, although the value and accuracy of both are sharply contested. The first, developed by Lawrence Farwell in the 1980s, is known as "brain fingerprinting." Subjects put on an electrode-filled helmet that measures a brain wave called p300, which, according to Farwell, changes its frequency when people recognize images, pictures, sights, and smells. After showing a suspect pictures of familiar places and measuring his p300 activation patterns, government officials could, at least in theory, show a suspect pictures of places he may or may not have seen before—an al Qaeda training camp, for example, or a crime scene—and compare the activation patterns. (By detecting not only lies but also honest cases of forgetfulness, the technology could expand our very idea of lie detection.)

The second lie-detection technology uses fMRI machines to compare the brain activity of liars and truth tellers. It is based on a test called Guilty Knowledge, developed by Daniel Langleben at the University of Pennsylvania in 2001. Langleben gave subjects a playing card before they entered the magnet and told them to answer "no" to a series of questions, including whether they had the card in question. Langleben and his colleagues found that certain areas of the brain lit up when people lied.

Two companies, No Lie MRI and Cephos, are now competing to refine fMRI lie-detection technology so that it can be admitted in court and commercially marketed. I talked to Steven Laken, the president of Cephos, which plans to begin selling its products this year. "We have two to three people who call every single week," he told me. "They're in legal proceedings throughout the world, and they're looking to bolster their credibility." Laken said the technology could have "tremendous applications" in civil and criminal cases. On the government side, he said, the technology could replace highly inaccurate polygraphs in screening for security clearances, as well as in trying to identify suspected terrorists' native languages and close associates. "In lab studies, we've been in the 80 to 90 percent accuracy range," Laken says. This is similar to the accuracy rate for polygraphs, which are not considered sufficiently reliable to be allowed in most legal cases. Laken says he hopes to reach the 90 percent to 95 percent accuracy range—which should be high enough to satisfy the Supreme Court's standards for the admission of scientific evidence. Judy Illes, director of neuroethics at the Stanford Center for Biomedical Ethics, says, "I would predict that within five years, we will have technology that is sufficiently reliable at getting at the binary question of whether someone is lying that it may be utilized in certain legal settings."

If and when lie-detection fMRIs are admitted in court, they will raise vexing questions of self-incrimination and privacy. Hank Greely, a law professor and head of the Stanford Center for Law and the Biosciences, notes that prosecution and defense witnesses might have their credibility questioned if they refused to take a lie-detection fMRI, as might parties and witnesses in civil cases. Unless courts found the tests to be shocking invasions of privacy, like stomach pumps, witnesses could even be compelled to have their brains scanned. And equally vexing legal questions might arise as neuroimaging technologies move beyond telling whether or not someone is lying and begin to identify the actual content of memories. Michael Gazzaniga, a professor of psychology at the University of California, Santa Barbara, and author of *The Ethical Brain,* notes that within 10 years, neuroscientists may be able to show that there are neurological differences when people testify about their own previous acts and when they testify to something they saw. "If you kill someone, you have a procedural memory of that, whereas if I'm standing and watch you kill somebody, that's an episodic memory that uses a different part of the brain," he told me. Even if witnesses don't have their brains scanned, neuroscience may lead judges and jurors to conclude that certain kinds of memories are more reliable than others because of the area of the brain in which they are processed. Further into the future, and closer to science fiction, lies the possibility of memory downloading. "One could even, just barely, imagine a technology that might be able to 'read out' the witness's memories, intercepted as neuronal firings, and translate it directly into voice, text or the equivalent of a movie," Hank Greely writes.

Greely acknowledges that lie-detection and memory-retrieval technologies like this could pose a serious challenge to our freedom of thought, which is now defended largely by the First Amendment protections for freedom of expression. "Freedom of thought has always been buttressed by the reality that you could only tell what someone thought based on their behavior," he told me. "This technology holds out the possibility of looking through the skull and seeing what's really happening, seeing the thoughts themselves." According to Greely, this may challenge the principle that we should be held accountable for what we do, not what we think. "It opens up for the first time the possibility of punishing people for their thoughts rather than their actions," he says. "One reason thought has been free in the harshest dictatorships is that dictators haven't been able to detect it." He adds, "Now they may be able to, putting greater pressure on legal constraints against government interference with freedom of thought."

In the future, neuroscience could also revolutionize the way jurors are selected. Steven Laken, the president of Cephos, says that jury consultants might seek to put prospective jurors in fMRIs. "You could give videotapes of the lawyers and witnesses to people when they're in the magnet and see what parts of their brains light up," he says. A situation like this would raise vexing questions about jurors' prejudices—and what makes for a fair trial. Recent experiments have suggested that people who believe themselves to be free of bias may harbor plenty of it all the same.

The experiments, conducted by Elizabeth Phelps, who teaches psychology at New York University, combine brain scans with a behavioral test known as the Implicit Association Test, or IAT, as well as physiological tests of the startle reflex. The IAT flashes pictures of black and white faces at you and asks you to associate various adjectives with the faces. Repeated tests have shown that white subjects take longer to respond when they're asked to associate black faces with positive adjectives and white faces with negative adjectives than vice versa, and this is said to be an implicit measure of unconscious racism. Phelps and her colleagues added neurological evidence to this insight by scanning the brains and testing the startle reflexes of white undergraduates at Yale before they took the IAT. She found that the subjects who showed the most unconscious bias on the IAT also had the highest activation in their amygdalas—a center of threat perception—when unfamiliar black faces were flashed at them in the scanner. By contrast, when subjects were shown pictures of familiar black and white figures—like Denzel Washington, Martin Luther King Jr., and Conan O'Brien—there was no jump in amygdala activity.

The legal implications of the new experiments involving bias and neuroscience are hotly disputed. Mahzarin R. Banaji, a psychology professor at Harvard who helped to pioneer the IAT, has argued that there may be a big gap between the concept of intentional bias embedded in law and the reality of unconscious racism revealed by science. When the gap is "substantial," she and UCLA law professor Jerry Kang have argued, "the law should be changed to comport with science"—relaxing, for example, the current focus on intentional discrimination and trying to root out unconscious bias in the workplace with "structural interventions," which critics say may be tantamount to racial quotas. One legal scholar has cited Phelps's work to argue for the elimination of peremptory challenges to prospective jurors—if most whites are unconsciously racist, the argument goes, then any decision to strike a black juror must be infected with racism. Much to her displeasure, Phelps's work has been cited by a journalist to suggest that a white cop who accidentally shot a black teenager on a Brooklyn rooftop in 2004 must have been responding to a hard-wired fear of unfamiliar black faces—a version of the amygdala made me do it.

Phelps herself says it's "crazy" to link her work to cops who shoot on the job and insists that it is too early to use her research in the courtroom. "Part of my discomfort is that we haven't linked what we see in the amygdala or any other region of the brain with an activity outside the magnet that we would call racism," she told me. "We have no evidence whatsoever that activity in the brain is more predictive of things we care about in the courtroom than the behaviors themselves that we correlate with brain function." In other words, just because you have a biased reaction to a photograph doesn't mean you'll act on those biases in the workplace. Phelps is also concerned that jurors might be unduly influenced by attention-grabbing pictures of brain scans. "Frank Keil, a psychologist at Yale, has done research suggesting that when you have a picture of a mechanism, you have a tendency to overestimate how much you understand the mechanism," she told me. Defense lawyers confirm this phenomenon. "Here was this nice color image we could enlarge, that the medical expert could point to," Christopher

Plourd, a San Diego criminal defense lawyer, told the *Los Angeles Times* in the early 1990s. "It documented that this guy had a rotten spot in his brain. The jury glommed onto that."

Other scholars are even sharper critics of efforts to use scientific experiments about unconscious bias to transform the law. "I regard that as an extraordinary claim that you could screen potential jurors or judges for bias; it's mind-boggling," I was told by Philip Tetlock, professor at the Haas School of Business at the University of California at Berkley. Tetlock has argued that split-second associations between images of African Americans and negative adjectives may reflect "simple awareness of the social reality" that "some groups are more disadvantaged than others." He has also written that, according to psychologists, "there is virtually no published research showing a systematic link between racist attitudes, overt or subconscious, and real-world discrimination." (A few studies show, Tetlock acknowledges, that openly biased white people sometimes sit closer to whites than blacks in experiments that simulate job hiring and promotion.) "A light bulb going off in your brain means nothing unless it's correlated with a particular output, and the brain-scan stuff, heaven help us, we have barely linked that with anything," agrees Tetlock's coauthor, Amy Wax of the University of Pennsylvania Law School. "The claim that homeless people light up your amygdala more and your frontal cortex less and we can infer that you will systematically dehumanize homeless people—that's piffle."

V. ARE YOU RESPONSIBLE FOR WHAT YOU MIGHT DO?

The attempt to link unconscious bias to actual acts of discrimination may be dubious. But are there other ways to look inside the brain and make predictions about an individual's future behavior? And, if so, should those discoveries be employed to make us safer? Efforts to use science to predict criminal behavior have a disreputable history. In the 19th century, the Italian criminologist Cesare Lombroso championed a theory of "biological criminality," which held that criminals could be identified by physical characteristics, like large jaws or bushy eyebrows. Nevertheless, neuroscientists are trying to find the factors in the brain associated with violence. PET scans of convicted murderers were first studied in the late 1980s by Adrian Raine, a professor of psychology at the University of Southern California; he found that their prefrontal cortexes, areas associated with inhibition, had reduced glucose metabolism and suggested that this might be responsible for their violent behavior. In a later study, Raine found that subjects who received a diagnosis of antisocial personality disorder, which correlates with violent behavior, had 11 percent less gray matter in their prefrontal cortexes than control groups of healthy subjects and substance abusers. His current research uses fMRIs to study moral decision making in psychopaths.

Neuroscience, it seems, points two ways: it can absolve individuals of responsibility for acts they've committed, but it can also place individuals in jeopardy for acts they haven't committed—but might someday. "This opens up a Pandora's box in civilized society that I'm willing to fight against," says Helen S. Mayberg, a professor of psychiatry, behavioral sciences, and neurology at Emory University School of Medicine, who has testified against the admission of neuroscience evidence in criminal trials. "If you believe at the time of trial that the picture informs us about what they were like at the time of the crime, then the picture moves forward. You need to be prepared for: 'This spot is a sign of future dangerousness,' when someone is up for parole. They have a scan, the spot is there, so they don't get out. It's carved in your brain."

Other scholars see little wrong with using brain scans to predict violent tendencies and sexual predilections—as long as the scans are used within limits. "It's not necessarily the case that if predictions work, you would say take that guy off the street and throw away the key," says Hank Greely, the Stanford law professor. "You could require counseling, surveillance, GPS transmitters, or warning the neighbors. None of these are necessarily benign, but they beat the heck out of preventative detention." Greely has little doubt that predictive technologies will be enlisted in the war on terror—perhaps in radical ways. "Even with today's knowledge, I think we can tell whether someone has a strong emotional reaction to seeing things, and I can certainly imagine a friend-versus-foe scanner. If you put everyone who reacts badly to an American flag in a concentration camp or Guantánamo, that would be bad, but in an occupation situation, to mark someone down for further surveillance, that might be appropriate."

Paul Root Wolpe, who teaches social psychiatry and psychiatric ethics at the University of Pennsylvania School of Medicine, says he anticipates that neuroscience predictions will move beyond the courtroom and will be used to make predictions about citizens in all walks of life.

"Will we use brain imaging to track kids in school because we've discovered that certain brain function or morphology suggests aptitude?" he asks. "I work for NASA, and imagine how helpful it might be for NASA if it could scan your brain to discover whether you have a good enough spatial sense to be a pilot." Wolpe says that brain imaging might eventually be used to decide whether someone is a worthy foster or adoptive parent—a history of major depression and cocaine abuse can leave telltale signs on the brain, for example, and future studies might find parts of the brain that correspond to nurturing and caring.

The idea of holding people accountable for their predispositions rather than their actions poses a challenge to one of the central principles of Anglo-American jurisprudence: namely, that people are responsible for their behavior, not their proclivities—for what they do, not what they think. "We're going to have to make a decision about the skull as a privacy domain," Wolpe says. Indeed, Wolpe serves on the board of an organization called the Center for Cognitive Liberty and Ethics, a group of neuroscientists, legal scholars, and privacy advocates "dedicated to protecting and advancing freedom of thought in the modern world of accelerating neurotechnologies."

There may be similar "cognitive liberty" battles over efforts to repair or enhance broken brains. A remarkable technique called transcranial magnetic stimulation, for example, has been used to stimulate or inhibit specific regions of the brain. It can temporarily alter how we think and feel. Using TMS, Ernst Fehr and Daria Knoch of the University of Zurich temporarily disrupted each side of the dorsolateral prefrontal cortex in test subjects. They asked their subjects to participate in an experiment that economists call the ultimatum game. One person is given $20 and told to divide it with a partner. If the partner rejects the proposed amount as too low, neither person gets any money. Subjects whose prefrontal cortexes were functioning properly tended to reject offers of $4 or less: they would rather get no money than accept an offer that struck them as insulting and unfair. But subjects whose right prefrontal cortexes were suppressed by TMS tended to accept the $4 offer. Although the offer still struck them as insulting, they were able to suppress their indignation and to pursue the selfishly rational conclusion that a low offer is better than nothing.

Some neuroscientists believe that TMS may be used in the future to enforce a vision of therapeutic justice, based on the idea that defective brains can be cured. "Maybe somewhere down the line, a badly damaged brain would be viewed as something that can heal, like a broken leg that needs to be repaired," the neurobiologist Robert Sapolsky says, although he acknowledges that defining what counts as a normal brain is politically and scientifically fraught. Indeed, efforts to identify normal and abnormal brains have been responsible for some of the darkest movements in the history of science and technology, from phrenology to eugenics. "How far are we willing to go to use neurotechnology to change people's brains we consider disordered?" Wolpe asks. "We might find a part of the brain that seems to be malfunctioning, like a discrete part of the brain operative in violent or sexually predatory behavior, and then turn off or inhibit that behavior using transcranial magnetic stimulation." Even behaviors in the normal range might be fine-tuned by TMS: jurors, for example, could be made more emotional or more deliberative with magnetic interventions. Mark George, an adviser to the Cephos company and also director of the Medical University of South Carolina Center for Advanced Imaging Research, has submitted a patent application for a TMS procedure that supposedly suppresses the area of the brain involved in lying and makes a person less capable of not telling the truth.

As the new technologies proliferate, even the neurolaw experts themselves have only begun to think about the questions that lie ahead. Can the police get a search warrant for someone's brain? Should the Fourth Amendment protect our minds in the same way that it protects our houses? Can courts order tests of suspects' memories to determine whether they are gang members or police informers, or would this violate the Fifth Amendment's ban on compulsory self-incrimination? Would punishing people for their thoughts rather than for their actions violate the Eighth Amendment's ban on cruel and unusual punishment? However astonishing our machines may become, they cannot tell us how to answer these perplexing questions. We must instead look to our own powers of reasoning and intuition, relatively primitive as they may be. As Stephen Morse puts it, neuroscience itself can never identify the mysterious point at which people should be excused from responsibility for their actions because they are not able, in some sense, to control themselves. That question, he suggests, is "moral and ultimately legal," and it must be answered not in laboratories but in courtrooms and legislatures. In other words, we must answer it ourselves.

Caleb Crain

Twilight of the Books

What will life be like if people stop reading?

In 1937, 29 percent of American adults told the pollster George Gallup that they were reading a book. In 1955, only 17 percent said they were. Pollsters began asking the question with more latitude. In 1978, a survey found that 55 percent of respondents had read a book in the previous six months. The question was even looser in 1998 and 2002, when the General Social Survey found that roughly 70 percent of Americans had read a novel, a short story, a poem, or a play in the preceding 12 months. And, this August, 73 percent of respondents to another poll said that they had read a book of some kind, not excluding those read for work or school, in the past year. If you didn't read the fine print, you might think that reading was on the rise.

You wouldn't think so, however, if you consulted the Census Bureau and the National Endowment for the Arts, who, since 1982, have asked thousands of Americans questions about reading that are not only detailed but consistent. The results, first reported by the NEA in 2004, are dispiriting. In 1982, 56.9 percent of Americans had read a work of creative literature in the previous 12 months. The proportion fell to 54 percent in 1992 and to 46.7 percent in 2002. In

November, the NEA released a follow-up report, "To Read or Not to Read," which showed correlations between the decline of reading and social phenomena as diverse as income disparity, exercise, and voting. In his introduction, the NEA chairman, Dana Gioia, wrote, "Poor reading skills correlate heavily with lack of employment, lower wages, and fewer opportunities for advancement."

This decline is not news to those who depend on print for a living. In 1970, according to *Editor and Publisher International Year Book,* there were 62.1 million weekday newspapers in circulation—about 0.3 papers per person. Since 1990, circulation has declined steadily, and in 2006 there were just 52.3 million weekday papers—about 0.17 per person. In January 1994, 49 percent of respondents told the Pew Research Center for the People and the Press that they had read a newspaper the day before. In 2006, only 43 percent said so, including those who read online. Book sales, meanwhile, have stagnated. The Book Industry Study Group estimates that sales fell from 8.27 books per person in 2001 to 7.93 in 2006. According to the Department of Labor, American households spent an average of $163 on reading in 1995 and $126 in 2005. In "To Read or Not to Read," the NEA reports that American households' spending on books, adjusted for inflation, is "near its twenty-year low," even as the average price of a new book has increased.

More alarming are indications that Americans are losing not just the will to read but even the ability. According to the Department of Education, between 1992 and 2003 the average adult's skill in reading prose slipped 1 point on a 500-point scale, and the proportion who were proficient— capable of such tasks as "comparing viewpoints in two editorials"—declined from 15 percent to 13. The Department of Education found that reading skills have improved moderately among 4th and 8th graders in the past decade and a half, with the largest jump occurring just before the No Child Left Behind Act took effect, but 12th graders seem to be taking after their elders. Their reading scores fell an average of 6 points between 1992 and 2005, and the share of proficient 12th grade readers dropped from 40 percent to 35 percent. The steepest declines were in "reading for literary experience"—the kind that involves "exploring themes, events, characters, settings, and the language of literary works," in the words of the department's test makers. In 1992, 44 percent of 12th graders told the Department of Education that they talked about their reading with friends at least once a week. By 2005, only 37 percent said they did.

The erosion isn't unique to America. Some of the best data come from the Netherlands, where in 1955 researchers began to ask people to keep diaries of how they spent every 15 minutes of their leisure time. Time-budget diaries yield richer data than surveys, and people are thought to be less likely to lie about their accomplishments if they have to do it four times an hour. Between 1955 and 1975, the decades when television was being introduced into the Netherlands, reading on weekday evenings and weekends fell from 5 hours a week to 3.6, while television watching rose from about 10 minutes a week to more than 10 hours. During the next two decades, reading continued to fall and television watching to rise, though more slowly. By 1995, reading, which had occupied 21 percent of people's spare time in 1955, accounted for just 9 percent.

The most striking results were generational. In general, older Dutch people read more. It would be natural to infer from this that each generation reads more as it ages, and, indeed, the researchers found something like this to be the case for earlier generations. But, with later ones, the age-related growth in reading dwindled. The turning point seems to have come with the generation born in the 1940s. By 1995, a Dutch college graduate born after 1969 was likely to spend fewer hours reading each week than a little-educated person born before 1950. As far as reading habits were concerned, academic credentials mattered less than whether a person had been raised in the era of television. The NEA, in its 20 years of data, has found a similar pattern. Between 1982 and 2002, the percentage of Americans who read literature declined not only in every age group but in every generation—even in those moving from youth into middle age, which is often considered the most fertile time of life for reading. We are reading less as we age, and we are reading less than people who were our age 10 or 20 years ago.

There's no reason to think that reading and writing are about to become extinct, but some sociologists speculate that reading books for pleasure will one day be the province of a special "reading class," much as it was before the arrival of mass literacy, in the second half of the 19th century. They warn that it probably won't regain the prestige of exclusivity; it may just become "an increasingly arcane hobby." Such a shift would change the texture of society. If one person decides to watch *The Sopranos* rather than to read Leonardo Sciascia's novella *To Each His Own,* the culture goes on largely as before—both viewer and reader are entertaining themselves while learning something about the Mafia in the bargain. But if, over time, many people choose television over books, then a nation's conversation with itself is likely to change. A reader learns about the world and imagines it differently from the way a viewer does; according to some experimental psychologists, a reader and a viewer even think differently. If the eclipse of reading continues, the alteration is likely to matter in ways that aren't foreseeable.

Taking the long view, it's not the neglect of reading that has to be explained but the fact that we read at all. "The act of reading is not natural," Maryanne Wolf writes in *Proust and the Squid,* an account of the history and biology of reading. Humans started reading far too recently for any of our genes to code for it specifically. We can do it only because the brain's plasticity enables the repurposing of circuitry that originally evolved for other tasks—distinguishing at a glance a garter snake from a haricot vert, say.

The squid of Wolf's title represents the neurobiological approach to the study of reading. Bigger cells are easier for scientists to experiment on, and some species of squid have optic-nerve cells a hundred times as thick as mammal neurons and up to four inches long, making them a favorite with biologists. (Two decades ago, I had a summer job washing glassware in Cape Cod's Marine Biological Laboratory. Whenever researchers extracted an optic nerve, they threw the rest of the squid into a freezer, and about once a month we took a cooler-full to the beach for grilling.) To symbolize the humanistic approach to reading, Wolf has chosen Proust, who described reading as "that fruitful miracle of a communication in the midst of solitude." Perhaps inspired by Proust's example, Wolf, a dyslexia researcher at Tufts, reminisces about the nuns who taught her to read in a two-room brick schoolhouse in Illinois. But she's more of a squid person than a Proust person and seems most at home when dissecting Proust's fruitful miracle into such brain parts as the occipital "visual association area" and "area 37's fusiform gyrus." Given the panic that takes hold of humanists when the decline of reading is discussed, her cold-blooded perspective is opportune.

Wolf recounts the early history of reading, speculating about developments in brain wiring as she goes. For example, from the eighth to the fifth millennia BCE, clay tokens were used in Mesopotamia for tallying livestock and other goods. Wolf suggests that, once the simple markings on the tokens were understood not merely as squiggles but as representations of, say, 10 sheep, they would have put more of the brain to work. She draws on recent research with functional magnetic resonance imaging (fMRI), a technique that maps blood flow in the brain during a given task, to show that meaningful squiggles activate not only the occipital regions responsible for vision but also temporal and parietal regions associated with language and computation. If a particular squiggle was repeated on a number of tokens, a group of nerves might start to specialize in recognizing it, and other nerves to specialize in connecting to language centers that handled its meaning.

In the fourth millennium BCE, the Sumerians developed cuneiform, and the Egyptians hieroglyphs. Both scripts began with pictures of things, such as a beetle or a hand, and then some of these symbols developed more abstract meanings, representing ideas in some cases and sounds in others. Readers had to recognize hundreds of symbols, some of which could stand for either a word or a sound, an ambiguity that probably slowed down decoding. Under this heavy cognitive burden, Wolf imagines, the Sumerian reader's brain would have behaved the way modern brains do when reading Chinese, which also mixes phonetic and ideographic elements and seems to stimulate brain activity in a pattern distinct from that of people reading the Roman alphabet. Frontal regions associated with muscle memory would probably also have gone to work, because the Sumerians learned their characters by writing them over and over, as the Chinese do today.

Complex scripts like Sumerian and Egyptian were written only by scribal elites. A major breakthrough occurred around 750 BCE, when the Greeks, borrowing characters from a Semitic language, perhaps Phoenician, developed a writing system that had just 24 letters. There had been scripts with a limited number of characters before, as there had been consonants and even occasionally vowels, but the Greek alphabet was the first whose letters recorded every significant sound element in a spoken language in a one-to-one correspondence, give or take a few diphthongs. In ancient Greek, if you knew how to pronounce a word, you knew how to spell it, and you could sound out almost any word you saw, even if you'd never heard it before. Children learned to read and write Greek in about three years, somewhat faster than modern children learn English, whose alphabet is more ambiguous. The ease democratized literacy; the ability to read and write spread to citizens who didn't specialize in it. The classicist Eric A. Havelock believed that the alphabet changed "the character of the Greek consciousness."

Wolf doesn't quite second that claim. She points out that it is possible to read efficiently a script that combines ideograms and phonetic elements, something that many Chinese do daily. The alphabet, she suggests, entailed not a qualitative difference but an accumulation of small quantitative ones, by helping more readers reach efficiency sooner. "The efficient reading brain," she writes, "quite literally has more time to think." Whether that development sparked Greece's flowering she leaves to classicists to debate, but she agrees with Havelock that writing was probably a contributive factor, because it freed the Greeks from the necessity of keeping their whole culture, including the *Iliad* and the *Odyssey,* memorized.

The scholar Walter J. Ong once speculated that television and similar media are taking us into an era of "secondary orality," akin to the primary orality that existed before the emergence of text. If so, it is worth trying to understand how different primary orality must have been from our own mind-set. Havelock theorized that, in ancient Greece, the effort required to preserve knowledge colored everything. In Plato's day, the word *mimesis* referred to an actor's performance of his role, an audience's identification with a performance, a pupil's recitation of his lesson, and an apprentice's emulation of his master. Plato, who was literate, worried about the kind of trance or emotional enthrallment that came over people in all these situations, and Havelock inferred from this that the idea of distinguishing the knower from the known was then still a novelty. In a society that had only recently learned to take notes, learning something still meant abandoning yourself to it. "Enormous powers of poetic memorization could be purchased only at the cost of total loss of objectivity," he wrote.

It's difficult to prove that oral and literate people think differently; orality, Havelock observed, doesn't "fossilize" except through its nemesis, writing. But some supporting evidence came to hand in 1974, when Aleksandr R. Luria, a Soviet psychologist, published a study based on interviews conducted in the 1930s with illiterate and newly literate peasants in Uzbekistan and Kyrgyzstan. Luria found that illiterates had a "graphic-functional" way of thinking that seemed to vanish as they were schooled. In naming colors, for example, literate people said "dark blue" or "light yellow," but illiterates used metaphorical names like "liver," "peach," "decayed teeth," and "cotton in bloom." Literates saw optical illusions; illiterates sometimes didn't. Experimenters showed peasants drawings of a hammer, a saw, an axe, and a log and then asked them to choose the three items that were similar. Illiterates resisted, saying that all the items were useful. If pressed, they considered throwing out the hammer; the situation of chopping wood seemed more cogent to them than any conceptual category. One peasant, informed that someone had grouped the three tools together, discarding the log, replied, "Whoever told you that must have been crazy," and another suggested, "Probably he's got a lot of firewood." One frustrated experimenter showed a picture of three adults and a child and declared, "Now, clearly the child doesn't belong in this group," only to have a peasant answer:

Oh, but the boy must stay with the others! All three of them are working, you see, and if they have to keep running out to fetch things, they'll never get the job done, but the boy can do the running for them.

Illiterates also resisted giving definitions of words and refused to make logical inferences about hypothetical situations. Asked by Luria's staff about polar bears, a peasant grew testy: "What the cock knows how to do, he does. What I know, I say, and nothing beyond that!" The illiterates did not talk about themselves except in terms of their tangible possessions. "What can I say about my own heart?" one asked.

In the 1970s, the psychologists Sylvia Scribner and Michael Cole tried to replicate Luria's findings among the Vai, a rural people in Liberia. Since some Vai were illiterate, some were schooled in English, and others were literate in the Vai's own script, the researchers hoped to be able to distinguish cognitive changes caused by schooling from those caused specifically by literacy. They found that English schooling and English literacy improved the ability to talk about language and solve logic puzzles, as literacy had done with Luria's peasants. But literacy in Vai script improved performance on only a few language-related tasks. Scribner and Cole's modest conclusion—"Literacy makes some difference to some skills in some contexts"—convinced some people that the literate mind was not so different from the oral one after all. But others have objected that it was misguided to separate literacy from schooling, suggesting that cognitive changes came with the culture of literacy rather than with the mere fact of it. Also, the Vai script, a syllabary with more than 200 characters, offered nothing like the cognitive efficiency that Havelock ascribed to Greek. Reading Vai, Scribner and Cole admitted, was "a complex problem-solving process," usually performed slowly.

Soon after this study, Ong synthesized existing research into a vivid picture of the oral mind-set. Whereas literates can rotate concepts in their minds abstractly, orals embed their thoughts in stories. According to Ong, the best way to preserve ideas in the absence of writing is to "think memorable thoughts," whose zing ensures their transmission. In an oral culture, cliché and stereotype are valued, as accumulations of wisdom, and analysis is frowned upon, for putting those accumulations at risk. There's no such concept as plagiarism, and redundancy is an asset that helps an audience follow a complex argument. Opponents in struggle are more memorable than calm and abstract investigations, so bards revel in name calling and in "enthusiastic description of physical violence." Since there's no way to erase a mistake invisibly, as one may in writing, speakers tend not to correct themselves at all. Words have their present meanings but no older ones, and if the past seems to tell a story with values different from current ones, it is either forgotten or silently adjusted. As the scholars Jack Goody and Ian Watt observed, it is only in a literate culture that the past's inconsistencies have to be accounted for, a process that encourages skepticism and forces history to diverge from myth.

Upon reaching classical Greece, Wolf abandons history, because the Greeks' alphabet-reading brains probably resembled ours, which can be readily put into scanners. Drawing on recent imaging studies, she explains in detail how a modern child's brain wires itself for literacy. The ground is laid in preschool, when parents read to a child, talk with her, and encourage awareness of sound elements like rhyme and alliteration, perhaps with Mother Goose poems. Scans show that when a child first starts to read she has to use more of her brain than adults do. Broad regions light up in both hemispheres. As a child's neurons specialize in recognizing letters and become more efficient, the regions activated become smaller.

At some point, as a child progresses from decoding to fluent reading, the route of signals through her brain shifts. Instead of passing along a "dorsal route" through occipital, temporal, and parietal regions in both hemispheres, reading starts to move along a faster and more efficient "ventral route," which is confined to the left hemisphere. With the gain in time and the freed up brainpower, Wolf suggests, a fluent reader is able to integrate more of her own thoughts and feelings into her experience. "The secret at the heart of reading," Wolf writes, is "the time it frees for the brain to have thoughts deeper than those that came before." Imaging studies suggest that in many cases of dyslexia the right hemisphere never disengages, and reading remains effortful.

In a recent book claiming that television and video games are "making our minds sharper," the journalist Steven Johnson argues that since we value reading for "exercising the mind," we should value electronic media for offering a superior "cognitive workout." But, if Wolf's evidence is right, Johnson's metaphor of exercise is misguided. When reading goes well, Wolf suggests, it feels effortless, like drifting down a river rather than rowing up it. It makes you smarter because it leaves more of your brain alone. Ruskin once compared reading to a conversation with the wise and noble, and Proust corrected him. It's much better than that, Proust wrote. To read is "to receive a communication with another way of thinking, all the while remaining alone, that is, while continuing to enjoy the intellectual power that one has in solitude and that conversation dissipates immediately."

Wolf has little to say about the general decline of reading, and she doesn't much speculate about the function of the brain under the influence of television and newer media. But there is research suggesting that secondary orality and literacy don't mix. In a study published this year, experimenters varied the way that people took in a PowerPoint presentation about the country of Mali. Those who were allowed to read silently were more likely to agree with the statement "The presentation was interesting," and those who read along with an audiovisual commentary were more likely to agree with the statement "I did not learn anything from this presentation." The silent readers remembered more, too, a finding in line with a series of British studies in which people who read transcripts of television newscasts, political programs, advertisements, and science shows recalled more information than those who had watched the shows themselves.

The antagonism between words and moving images seems to start early. In August, scientists at the University of Washington revealed that babies between the ages of 8 and 16 months know on average six to eight fewer words for every hour of baby DVDs and videos they watch daily. A 2005 study in northern California found that a television in the bedroom lowered the standardized test scores of third graders. And the conflict continues throughout a child's development. In 2001, after analyzing data on more than a million students around the world, the researcher Micha Razel found "little room for doubt" that television worsened performance in reading, science, and math. The relationship wasn't a straight line but "an inverted check mark": a small amount of television seemed to benefit children; more hurt. For 9-year-olds, the optimum was two hours a day; for 17-year-olds, half an hour. Razel guessed that the younger children were watching educational shows, and, indeed, researchers have shown that a 5-year-old boy who watches *Sesame Street* is likely to have higher grades even in high school. Razel noted, however, that 55 percent of students were exceeding their optimal viewing time by three hours a day, thereby lowering their academic achievement by roughly one grade level.

The internet, happily, does not so far seem to be antagonistic to literacy. Researchers recently gave Michigan children and teenagers home computers in exchange for permission to monitor their internet use. The study found that grades and reading scores rose with the amount of time spent online. Even visits to pornography Web sites improved academic performance. Of course, such synergies may disappear if the internet continues its YouTube-fueled evolution away from print and toward television.

No effort of will is likely to make reading popular again. Children may be browbeaten, but adults resist interference with their pleasures. It may simply be the case that many Americans prefer to learn about the world and to entertain themselves with television and other streaming media rather than with the printed word and that it is taking a few generations for them to shed old habits like newspapers and novels. The alternative is that we are nearing the end of a pendulum swing and that reading will return, driven back by forces as complicated as those now driving it away.

But if the change is permanent, and especially if the slide continues, the world will feel different, even to those who still read. Because the change has been happening slowly for decades, everyone has a sense of what is at stake, though it is rarely put into words. There is something to gain, of course, or no one would ever put down a book and pick up a remote. Streaming media give actual pictures and sounds instead of mere descriptions of them. "Television completes the cycle of the human sensorium," Marshall McLuhan proclaimed in 1967. Moving and talking images are much richer in information about a performer's appearance, manner, and tone of voice, and they give us the impression that we know more about her health and mood, too. The viewer may not catch all the details of a candidate's health care plan, but he has a much more definite sense of her as a personality, and his response to her is therefore likely to be more full of emotion. There is nothing like this connection in print. A feeling for a writer never touches the fact of the writer herself, unless reader and writer happen to meet. In fact, from Shakespeare to Pynchon, the personalities of many writers have been mysterious.

Emotional responsiveness to streaming media harks back to the world of primary orality, and, as in Plato's day, the solidarity amounts almost to a mutual possession. "Electronic technology fosters and encourages unification and involvement," in McLuhan's words. The viewer feels at home with his show, or else he changes the channel. The closeness makes it hard to negotiate differences of opinion. It can be amusing to read a magazine whose principles you despise, but it is almost unbearable to watch such a television show. And so, in a culture of secondary orality, we may be less likely to spend time with ideas we disagree with.

Self-doubt, therefore, becomes less likely. In fact, doubt of any kind is rarer. It is easy to notice inconsistencies in two written accounts placed side by side. With text, it is even easy to keep track of differing levels of authority behind different pieces of information. The trust that a reader grants to the *New York Times,* for example, may vary sentence by sentence. A comparison of two video reports, on the other hand, is cumbersome. Forced to choose between conflicting stories on television, the viewer falls back on hunches or on what he believed before he started watching. Like the peasants studied by Luria, he thinks in terms of situations and story lines rather than abstractions.

And he may have even more trouble than Luria's peasants in seeing himself as others do. After all, there is no one looking back at the television viewer. He is alone, though he, and his brain, may be too distracted to notice it. The reader is also alone, but the NEA reports that readers are more likely than non-readers to play sports, exercise, visit art museums, attend theater, paint, go to music events, take photographs, and volunteer. Proficient readers are also more likely to vote. Perhaps readers venture so readily outside because what they experience in solitude gives them confidence. Perhaps reading is a prototype of independence. No matter how much one worships an author, Proust wrote, "all he can do is give us desires." Reading somehow gives us the boldness to act on them. Such a habit might be quite dangerous for a democracy to lose.

Robin Mejia

These Images Document an Atrocity

Commercial satellites may be providing a power-
ful new weapon in the struggle to stop genocide.

On the corner of Jeremy Nelson's L-shaped desk at
Amnesty International's Washington office sit two 17-inch
monitors. Both have an extra-tall base, and even then they
rest on phone books to get them close to eye level for the six-
foot-three researcher. At 7:00 on a Thursday night in April,
an exhausted but upbeat Nelson is staring at two satellite
images of the same area in South Darfur, Sudan, one on each
screen. One was taken in December 2004 and the other in
February 2007. They show a region that was targeted by
what Sudan's government called a "road-clearing offensive." Amnesty reports indicate that last November, while
officials were engaged in peace talks in Nigeria, military
ground and air forces and Janjaweed militias burned dozens
of villages.

This is just one incident in a violent conflict that has
killed between 200,000 and 450,000 people in the Darfur region since 2003. A major obstacle to international interven-
tion has been the Sudanese government's refusal to acknowledge the level of violence and its own complicity. In
late March, for example, Sudanese president Omar Hassan

al-Bashir said in a TV interview that the U.S. State Department map showing 1,000 Darfurian villages as burned was a fabrication. The Sudanese government also holds that only 9,000 people have died in the bloodshed and that local Janjaweed militias—the same Janjaweed that gained notoriety for atrocities in Sudan's recent civil war—are independent actors, despite the fact that they've been seen attacking with military support, raping women and girls, pillaging and sometimes burning entire villages to the ground. Amnesty had been documenting the violence, but last year, the government stopped letting Amnesty's researchers into the country. Then last month, Sudan cited lack of evidence in refusing to comply with the International Criminal Court's arrest warrants for the minister of state for humanitarian affairs and a Janjaweed militia leader.

Which is where Nelson comes in. The 31-year-old researcher is an associate with Amnesty's Crisis Prevention and Response Center. Normally, he tracks hot spots and brewing crises, handles logistics, and develops graphics for postcard and poster campaigns. Now he's also learning to analyze satellite images. The catalyst is a partnership between Amnesty and the American Association for the Advancement of Science (AAAS) that's pioneering a new kind of human rights observation: the use of high-resolution satellite imagery—commercially available only since 2001—to document atrocities in areas made inaccessible to watchdog groups. The unusual collaboration started about a year ago, with test projects looking at Zimbabwe and Lebanon. The Darfur effort is by far their biggest yet and the most politically significant.

Since 2004, the African Union has maintained a modest force of about 7,000 peacekeepers in Darfur, but their mandate expires at the end of this month. The AAAS/Amnesty group hopes that the satellite images it is collecting will provide incontrovertible proof of burning and destruction. Ideally, Sudan then will be forced to accept the United Nations peacekeeping force that the government refused to allow in last year.

"What this satellite technology does, it makes it possible to break down those walls of secrecy. Not only to get information, but to get information in a way that's irrefutable," says Larry Cox, executive director of Amnesty International USA.

The images on Nelson's screen were taken by Quick-Bird, a satellite launched in 2001 by Colorado-based Digital-Globe, one of two U.S.-based commercial satellite companies. QuickBird's resolution is good enough to show individual houses and sometimes even cars, and it shoots in color. The other U.S. company, Dulles-based GeoEye, operates two similar satellites with slightly lower resolution. Each of the satellites orbits the planet several times a day; among the three, they reach almost any spot on Earth about once a week. It's fair to assume that government spy satellites still have the best equipment in orbit. But today, anyone with a big enough checkbook can order spy-quality images (with some exceptions: a 1997 U.S. law prohibits the collection and release of satellite imagery of Israel with a resolution better than two meters, for example).

Tonight, Nelson begins his work by making a copy of the shot in the right-hand screen and pasting it directly over the one on the left. Then he makes the top one nearly transparent. A river that cuts through the scene becomes a marker to help him line up the two. Now he can easily flip back and forth to look for changes.

Sudanese huts tend to follow a similar pattern: a solid base ring with a steep, thatched roof. In the earlier image, they show up as small circles, with a slight shading to the dome, depending on the direction of the sun. Nelson draws a small, green circle slightly larger than the area of the average hut and makes several dozen copies of it. Then he begins methodically placing a green circle over every hut that can be found in any of the half dozen settlements spread across the desert landscape.

When he finishes, he moves the 2007 shot to the top and begins the analysis again. When the roof of a thatched hut burns, the base often survives, leaving a telltale ring. But parts of this region were burned so thoroughly that there's nothing left but a large black scar. If you didn't know that huts were there before, you'd have no idea they were now gone.

"Whoever did this did a good job," he says quietly. "Thorough, at least."

By 8:00 p.m., he has a final tally: Out of 461 structures in the "before" image, most of them homes, 339 were destroyed, 5 were probably destroyed, and 117 were intact. He couldn't tell whether those 117 were still inhabited.

The origins of the Amnesty/AAAS collaboration date to October 2004, when Nelson's boss, Ariela Blätter, the director of Amnesty's Crisis Prevention and Response Center, got an invitation to a panel discussion on the crisis in Darfur. AAAS had invited speakers from the U.S. Agency for International Development and the State Department to discuss how they'd used high-resolution satellite imagery to map refugee flows. A few years before, the Amnesty group in Denmark had tried using satellite imagery to analyze fire patterns in Darfur, but the imagery had proved too ambiguous to have much of an impact. The State Department's shots, it turned out, were much clearer—and they came from a commercial satellite, presumably one anyone with funding could access.

"Is this something the human rights community could use?" Blätter remembers raising her hand to ask.

Lars Bromley, a tech-savvy AAAS geographer, was also in the audience. The then-29-year-old understood exactly what the State Department had done to create its report and what it would take for him and AAAS to do the same thing. All he needed was detailed knowledge of where problems were unfolding—the kind of information researchers at Amnesty could provide.

"He dragged me off of the conference for this intense conversation," recalls Blätter, now 33. "It quickly became clear that they had been looking for me and I had been looking for them."

Hoping to appear serious about technology herself, Blätter asked him if he'd seen Amnesty Denmark's study. "He told me it was cute," she says.

With Blätter's help, Bromley got funding from the MacArthur Foundation for a pilot project. Darfur was on both their minds, but it was too complicated a situation to try as a test case. So they decided to look at Zimbabwe. In the summer of 2005, President Robert Mugabe's government forces had razed settlements around the country, leaving thousands homeless. The areas targeted had been those that had voted heavily for the opposition, opposition leaders said. Amnesty and the United Nations issued a small mountain of reports describing the nature and scale of the destruction and documenting the government's subsequent denial of access to aid organizations. Still, the Zimbabwean government insisted that the operation was an urban renewal project with no political agenda.

Blätter and Bromley decided to get satellite images of the settlements before and after the razing. There were plenty of archival shots available for purchase, and, through his MacArthur grant, Bromley had money to commission new acquisitions. The only obstacle was getting exact locations: Amnesty has traditionally focused on the personal stories of eyewitnesses, not latitudes and longitudes. So the duo had to be resourceful. Amnesty's London researchers were able to map a settlement called Porta Farm by scanning Google Earth for a site fitting its description: on the main road out of Harare, going toward Bulawayo, between two lakes. For harder-to-find locations, Bromley created a map of the area and e-mailed it to local activists, who e-mailed coordinates back. Bromley placed the order and crossed his fingers.

"You sit there and wonder, 'Did I just waste X amount of dollars on images of the beautiful Zimbabwe countryside?'" he says.

But the shots were spot on. Porta Farm, for example, had consisted of more than 850 buildings that had housed at least 6,000 people. Now, except for dim traces of old dirt tracks, it was an empty landscape. In May 2006, Amnesty issued a news release and, with AAAS, put the images online. Their power quickly became clear. Even though the attacks had happened the previous year and were well documented, this publicity push generated more coverage for the situation in Zimbabwe—from outlets including the BBC and al-Jazeera—than Amnesty had in the previous 10 years. Amnesty's Zimbabwe campaign staff started giving interviews at 1:00 a.m., coinciding with the news release, and kept going till 10:00 that evening.

Then that summer, a group called Zimbabwe Lawyers for Human Rights, which through Zimbabwe's court system had been unsuccessfully fighting the forced evictions, submitted the images as evidence in a complaint filed before the African Commission on Human and Peoples' Rights. Mugabe's government seems to have been caught off-guard; officials requested a delay to have the images independently analyzed.

In the aftermath, both Amnesty and AAAS agreed enthusiastically to support the continued partnership, and the MacArthur Foundation funded an expansion. This grant covered Bromley's salary for the next three years. (Researchers at AAAS, like those at most scientific institutions, are expected to cover part or all of their salaries through research grants.) It also allowed Bromley to bring on interns to focus on new areas, including Darfur. A smaller grant, from the Open Society Institute, would pay for a look at Burma. Blätter got a grant from the Save Darfur Coalition and committed Jeremy Nelson to the project. She also started thinking about using the technology to explore other regions that Amnesty researchers couldn't physically reach. In Eritrea, for example, satellite images might help to locate makeshift jails deep in the interior, where she has heard that political prisoners are held in secret.

On Friday, Bromley, Nelson, and Blätter are on a conference call. The Amnesty International and AAAS offices are only a few minutes from each other, but this is how most of the group's work gets done.

This is the next phase of the project: They're creating an endangered list of sorts, villages Amnesty thinks are threatened in Darfur and across the border in Chad, where the conflict has spread. The hope is that satellite images of these towns will generate enough media attention to provide some protection.

"We'll let the Sudanese government know that we're watching," Blätter notes. "And asking the global community to join us."

Over the past week, London-based Julie Flint has been gathering intelligence. A researcher and independent journalist who has been working on, and sometimes in, Darfur since 1992, Flint has provided most of the location information the researchers have used for this project. Flint communicates with the Washington team mostly by e-mail, and she tends to send each thought as she has it. Nelson, Blätter, and Bromley are used to waking up to an in-box full of one- or two-sentence missives—locations where Janjaweed militias have been seen massing, updates on which towns one side or the other appears to be taking an interest in, word of where attacks are likely to occur soon. For today's call, Nelson has consolidated the bits on potential attacks into one document, giving the names and locations of towns Flint thinks are at risk.

Before they begin discussing the towns, Bromley notes that each image of Sudan might cost less than he'd expected. Bromley had ordered all of the "before" and many of the "after" images of attacked villages out of DigitalGlobe's archive catalog. But when he inquired about getting new imagery, he found out that QuickBird was booked solid over Sudan until well into summer.

Apparently, someone with deep pockets is very interested in Sudan, though whether it's government or private enterprise is impossible to know—DigitalGlobe doesn't release customer information.

There's an upside, however: Users don't purchase the satellite images they requisition; they license them. So after an image is delivered to the original buyer, it'll go into an archive. This system has provided the historical coverage that has allowed Amnesty and AAAS to collect "before" shots.

Because someone else is directing QuickBird, Bromley had to choose another satellite, one of GeoEye's, for the new shots of Darfur. Then, suddenly, that satellite died; apparently, a small component between the satellite's sensor and its memory failed. So Bromley located a third satellite, one run out of the Netherlands Antilles by a company called ImageSat. It can capture almost the same level of detail as QuickBird, but its cameras record in black and white. Blät-

ter and Nelson are worried that these images may not res-
onate with viewers as deeply as the color shots, but they're
the best the team can do for now.

That's when the possibility of a price reduction appeared. From a message Bromley just received, these images
of Sudan may cost about $1,600 each, $900 less than expected. (Satellite companies don't post their rates, but the
group has been paying about $2,500 a shot.) Odd—but
good—news, because at the end of the day, this project has a
budget. Blätter has $50,000, about enough for 20 shots, to
spend on the threatened villages. And as the point is to let
Sudan's government know that Amnesty can order more
anytime, she can't use up the whole budget immediately.
With this lower price, she'll be able to get more, but there's
just not enough money for every village on the initial list.

"So we have to make some hard decisions," she says, beginning the discussion. Bromley's looking at his list as she
talks. Like Blätter and Nelson, he hasn't slept much this
week. A half-eaten Pop-Tart is pushed off to the side of his
desk, and two Starbucks cups sit near his monitors.

"Kafod—no one has gotten in since 2006," Blätter notes.
In theory, at least, even oppressive governments don't want
their people to starve; they often let humanitarian aid
groups such as Oxfam operate where they don't welcome
Amnesty or Human Rights Watch. But aid agencies are occasionally kicked out or, more often, forced to evacuate because of threats to their workers' safety. (After Doctors
Without Borders released a report documenting rapes in a
Darfur refugee camp in 2005, Sudan issued an arrest warrant for the group's country chief.) Aid agencies haven't
been able to enter Kafod for about a year.

"I like it," Bromley says. "If we can get a current image,
it can be used by humanitarian organizations."

Bromley writes "yes" next to Kafod, and the discussion

moves on. Abu Sakim is a smallish settlement in North Darfur, not far from Kafod. It's under control of the Sudan Liberation Army, one of the main rebel groups, but it's just a few miles south of a Janjaweed stronghold, and militias are believed to have designs on it. Blätter points out that many people believe—mistakenly—that it's quiet in northern Sudan; a focus on Kafod and Abu Sakim might help change that perception. As a bonus, it looks as though the two towns might be close enough to each other to capture in a single shot.

Then there's Boldong, a town on the side of a fertile mountain that was attacked and burned early in the conflict but has since been rebuilt. A center for state-run logging operations before the war, Boldong is a strategic site. It's now held by the rebels, but according to Amnesty's information, the government wants to retake it, possibly to restart logging operations.

The team definitely wants Boldong.

Next up is Fanga. Near the same fertile soils as Boldong, the town has been attacked several times, but so far, it looks as though the Janjaweed and the government forces still haven't occupied it. Bromley notes that because of the multiple attacks, new destruction might not be as clear. Blätter says she's going to put parentheses around it.

Blätter sighs. "I want to get all of them," she says. But that's not an option. She pauses and then, in a businesslike tone, says she'll think some more before making a decision on Fanga.

And so it goes, down the list. Bromley stops at Silea. Flint had described a town of at least 500 households, but Google Earth shows only a few scatted huts. He asks Nelson to double-check the location.

Saturday afternoon, Bromley is back at his New York Avenue office. The night before, he discovered a typo in the list they'd drawn up—one that could have meant putting in a $2,500 order for the absolutely wrong location. This has been a rushed week. So today, he's going to redo the list from scratch, quietly and free of distractions, and make sure he gets the same coordinates.

Before he maps, he pulls up a program he calls his "fuzzy matcher." It's based on a set of village names and co-ordinates for Sudan generated by the United Nations. When he types in a name, the program searches for exact or, as the name suggests, fuzzy matches, returning all possibilities and their official coordinate information. He types in "Boldong."

"Now this is what you want to see," he says. "One 'Boldong' and where we expect to find it."

The phone rings. It's Nelson calling from across town. Over the past two days, both men repeatedly measured the distance between Abu Sakim and Kafod. It turns out the towns are about six miles apart. If the researchers put in a special, slightly more expensive order, they might get both in one shot. But there's also a chance the satellite would capture just one town at the very far edge of the frame or, even worse, that it would only get half of one town. The risk's just not worth it. Abu Sakim's out.

"That's tough math right there," Bromley says. "Hopefully, the information makes your decision for you. But sometimes you have to choose one over the other, and literally you're just looking at a Word document." (Days later, the team will learn how good a decision this was: Abu Sakim had made the list by mistake—by the time they were discussing it, it had already been overrun.)

Moving down his list, Bromley types "Hashaba" into the fuzzy matcher. On this one, there are 36 near-matches.

By late afternoon, Bromley has what looks like a solid coordinate list for all of the threatened villages, so he turns his attention to the images he has in hand. He's going to convert them into Google Earth files, viewable by the 100 million people or so who have downloaded the program.

He pulls one up. It could be an archaeological site: Small, round outlines mark where buildings once stood; old fence lines are still slightly visible as marks in the sand. With no context, the scene is beautiful. But with a little bit of background, it's devastating. Those small, round outlines recently were homes. The people who lived there, if still alive, are probably in refugee camps. You can see those in other images: tightly packed huts and tents that house tens of thousands of people in squalid conditions. In one image taken in February near a village called Tawila, the huts are built right up to the fence line of an African Union outpost, apparently in the hope of garnering a bit of extra protection.

Bromley stays home on Sunday. He can log into his work computers from there; he needs a good night's sleep and some time with his dog. Nelson isn't so lucky. He hasn't taken a full day off in more than two weeks, but his work computer isn't accessible from off-site, and he has another half dozen image pairs to analyze, so he's at the office again, though dressed down in jeans and a Nationals cap.

At midnight, Blätter calls to check in. "Oh, that's not good," she says when Nelson picks up. She's worried about the hours he's keeping and was hoping she'd reach his voice mail. He promises to call it a night.

By Monday afternoon, the team has accomplished a lot. Blätter has made the final decisions about the threatened sites and has reviewed a thick stack of documentation on the villages that have already been attacked. Julie Flint has just returned from Chad, where she interviewed refugees, collecting heartbreaking photos and testimonials that will humanize the satellite shots when Amnesty posts them online.

Nelson's reviewing the information that will be shipped to the graphic design firm responsible for the group's Web site.

With his part of the project under control, Bromley takes a break from Darfur. Just down the hall from his office is the AAAS Burma conflict-monitoring center. Intern Sean O'Connor, a 25-year-old who's starting grad school in the fall, has a desk that is separated from the hall by a tall counter. Bromley leans on this when he stops by to discuss a package they're getting ready to send to their contacts in Thailand.

O'Connor has been working with Bromley to document the Burmese government's persecution of the Karen, an ethnic minority group that lives along the country's mountainous border with Thailand. If anything, this task has proved even more challenging than documenting the destruction in Darfur: Rather than huts in a desert, the targets are homes in a jungle, in a part of the world often hidden by clouds. But the duo has documented a number of attacks and, a few weeks ago, posted early findings on Google Earth. The activists they'd been working with were impressed but afraid that the detailed information the images provided might help the military find local hideouts. So Bromley took the files down, and now they're mailing them on a DVD, along with a paper printout of a satellite image that could be taken to a refugee camp.

"This is called participatory mapping," Bromley says. "Rather than me speculating what things are, they can say, 'Look, there's my house.'"

As the week moves on, O'Connor gets information from the Free Burma Rangers that a one-day government offensive has just burned down four Burmese villages, leaving about 1,000 people homeless. The towns are small and close to each other, so he puts in a satellite order that should encompass all four. But the formal start of the monsoon season is less than two weeks away, and cloud cover is already becoming a problem. QuickBird may not get a clear image until next week, next month—or even in the fall, after the monsoons. By fall, any burn scars will have been overgrown by jungle. Two of the villages aren't in DigitalGlobe's library; unless new images are taken soon, there won't be evidence that they ever existed.

As O'Connor tracks the situation in Burma, the images for the at-risk Darfurian villages start to come in. It turns out that Bromley was mistaken about some images costing less than expected. But after Bromley explains the project, the company offers him a deal: For every 10 new images ordered, 2 additional ones will be free. The other companies Bromley works with have given him great deals on their archival images—one has even donated some—but this is the first break on new collections. It means the group will be able to afford to shoot most, though not all, of the threatened sites on their list.

As the first set of images from the Netherlands satellite arrives, Bromley encounters another glitch: Some aren't crisp enough. Satellite companies list the resolution their cameras get when they're pointing straight down at the ground, but satellites often end up shooting at a slight angle for speed. The larger the angle, the lower the resolution. Depending on why one wants an image, the difference may not matter. But Darfurian huts are small.

"If it's too fuzzy, we can't see if they're damaged," Bromley says.

Fortunately, the company agrees to redo the shots. The next week, those images arrive, and they're what he'd hoped for: crisp, unassailable evidence of villages that still exist— and proof to Sudan's government that someone is watching.

Michael Behar

The Prophet of Garbage

*Joseph Longo's Plasma Converter turns our most
vile and toxic trash into clean energy—and
promises to make a relic of the landfill.*

It sounds as if someone just dropped a tricycle into a meat
grinder. I'm sitting inside a narrow conference room at a research facility in Bristol, Connecticut, chatting with Joseph
Longo, the founder and CEO of Startech Environmental
Corporation. As we munch on takeout Subway sandwiches,
a plate-glass window is the only thing separating us from the
adjacent lab, which contains a glowing caldera of "plasma"
three times as hot as the surface of the sun. Every few minutes there's a horrific clanking noise—grinding followed by
a thunderous voomp, like the sound a gas barbecue makes
when it first ignites.

"Is it supposed to do that?" I ask Longo nervously.
"Yup," he says. "That's normal."

Despite his 74 years, Longo bears an unnerving resem-
blance to the longtime cover boy of *Mad* magazine, Alfred
E. Neuman, who shrugs off nuclear Armageddon with the
glib catchphrase "What, me worry?" Both share red hair, a
smattering of freckles, and a toothy grin. When such a man
tells me I'm perfectly safe from a 30,000°F arc of man-made

lightning heating a vat of plasma that his employees are "controlling" in the next room—well, I'm not completely reassured.

To put me at ease, Longo calls in David Lynch, who manages the demonstration facility. "There's no flame or fire inside. It's just electricity," Lynch assures me of the multimillion-dollar system that took Longo almost two decades to design and build. Then the two usher me into the lab, where the gleaming 15-foot-tall machine they've named the Plasma Converter stands in the center of the room. The entire thing takes up about as much space as a two-car garage, surprisingly compact for a machine that can consume nearly any type of waste—from dirty diapers to chemical weapons—by annihilating toxic materials in a process as old as the universe itself.

Called plasma gasification, it works a little like the big bang, only backward (you get nothing from something). Inside a sealed vessel made of stainless steel and filled with a stable gas—either pure nitrogen or, as in this case, ordinary air—a 650-volt current passing between two electrodes rips electrons from the air, converting the gas into plasma. Current flows continuously through this newly formed plasma, creating a field of extremely intense energy very much like lightning. The radiant energy of the plasma arc is so powerful it disintegrates trash into its constituent elements by tearing apart molecular bonds. The system is capable of breaking down pretty much anything except nuclear waste, the isotopes of which are indestructible. The only by-products are an obsidian-like glass used as a raw material for numerous applications, including bathroom tiles and high-strength asphalt, and a synthesis gas, or "syngas"—a mixture of primarily hydrogen and carbon monoxide that can be converted into a variety of marketable fuels, including ethanol, natural gas, and hydrogen.

Perhaps the most amazing part of the process is that it's self-sustaining. Just like your toaster, Startech's Plasma Converter draws its power from the electrical grid to get started. The initial voltage is about equal to the zap from a police stun gun. But once the cycle is under way, the 2,200°F syngas is fed into a cooling system, generating steam that drives turbines to produce electricity. About two-thirds of the power is siphoned off to run the converter; the rest can be used on-site for heating or electricity or sold back to the utility grid. "Even a blackout would not stop the operation of the facility," Longo says.

It all sounds far too good to be true. But the technology works. Over the past decade, half a dozen companies have been developing plasma technology to turn garbage into energy. "The best renewable energy is the one we complain about the most: municipal solid waste," says Louis Circeo, the director of plasma research at the Georgia Institute of Technology. "It will prove cheaper to take garbage to a plasma plant than it is to dump it on a landfill." A Startech machine that costs roughly $250 million could handle 2,000 tons of waste daily, approximately what a city of a million people amasses in that time span. Large municipalities typically haul their trash to landfills, where the operator charges a "tipping fee" to dump the waste. The national average is $35 a ton, although the cost can be more than twice that in the Northeast (where land is scarce, tipping fees are higher). And the tipping fee a city pays doesn't include the price of trucking the garbage often hundreds of miles to a landfill or the cost of capturing leaky methane—a greenhouse gas— from the decomposing waste. In a city with an average tipping fee, a $250-million converter could pay for itself in about 10 years, and that's without factoring in the money made from selling the excess electricity and syngas. After that break-even point, it's pure profit.

Someday very soon, cities might actually make money from garbage.

TALKING TRASH

It was a rainy morning when I pulled up to Startech R&D to see Longo waiting for me in the parking lot. Wearing a bright yellow oxford shirt, a striped tie, and blue pinstriped pants, he dashed across the blacktop to greet me as I stepped from my rental car. A street-smart Brooklyn native, Longo was an only child raised by parents who worked long hours at a local factory that made baseballs and footballs. He volunteered to fight in Korea as a paratrooper after a friend was killed in action. He's fond of antiquated slang like *attaboy* and *shills* (as in "those shills stole my patents") and is old-school enough to have only recently abandoned the protractors, pencils, and drafting tables that he used to design his original Plasma Converter in favor of computers.

Today, Longo is meeting with investors from U.S. Energy, a trio of veteran waste-disposal executives who recently formed a partnership to build the first plasma gasification plant on Long Island, New York. They own a transfer station (where garbage goes for sorting en route to landfills) and are in the process of buying six Startech converters to handle 3,000 tons of construction debris a day trucked from sites around the state. "It's mostly old tile, wood, nails, glass, metal, and wire all mixed together," one of the project's partners, Troy Caruso, tells me. For the demonstration, Longo prepares a sampling of typical garbage—bottles of leftover prescription drugs, bits of fiberglass insulation, a half-empty can of Slim-Fast. A conveyer belt feeds the trash into an auger, which shreds and crushes it into pea-size morsels (that explains the deafening grinding sound) before stuffing it into the plasma-reactor chamber.

The room is warm and humid, and a dull hum emanates from the machinery.

Caruso and his partners, Paul Marazzo and Michael Nuzzi, are silent at first. They've seen the demo before. But as more trash vanishes into the converter, they become increasingly animated, spouting off facts and figures about how the machine will revolutionize their business. "This technology eliminates the landfill, which is 80 percent of our costs," Nuzzi says. "And we can use it to generate fuel at the back end," adds Marazzo, who then asks Lynch if the converter can handle chunks of concrete (answer: yes). "The bottom line is that nobody wants a landfill in their backyard," Nuzzi tells me. New York City is already paying an astronomical $90 a ton to get rid of its trash. According to Startech, a few 2,000-ton-per-day plasma gasification plants could do it for $36. Sell the syngas and surplus electricity, and you'd actually net $15 a ton. "Gasification is not just environmentally friendly," Nuzzi says. "It's a good business decision."

The converter we're watching vaporize Slim-Fast is a mini version of Startech's technology, capable of consuming five tons a day of solid waste, or about what 2,200 Americans toss in the trash every 24 hours. Fueled with garbage from the local dump, the converter is fired up whenever Longo pitches visiting clients.

Longo has been talking with the National Science Foundation about installing a system at McMurdo Station in Antarctica. The Vietnamese government is considering buying one to get rid of stockpiles of Agent Orange that the U.S. military left behind after the war. Investors from China, Poland, Japan, Romania, Italy, Russia, Brazil, Venezuela, the United Kingdom, Mexico, and Canada have all entered contract negotiations with Startech after making the pilgrimage to Bristol to see Longo's dog-and-pony show.

Startech isn't the only company using plasma to turn waste into a source of clean energy. A handful of start-ups—Geoplasma, Recovered Energy, PyroGenesis, EnviroArc, and Plasco Energy, among others—have entered the market in the past decade. But Longo, who has worked in the garbage business for four decades, is perhaps the industry's most passionate founding father. "What's so devilishly wonderful about plasma gasification is that it's completely circular," he says. "It takes everything back to its fundamental components in a way that's beautiful." Although all plasma gasification systems recapture syngas to turn into fuel, Startech's "Starcell" system seems to be ahead of the pack in its ability to economically convert the substance into eco-friendly and competitively priced fuels. "A lot of other gasification technologies require multiple steps. This is a one-step process," says Patrick Davis of the U.S. Department of Energy's office of hydrogen production and delivery, which has awarded Longo's company almost $1 million in research grants. "You put the waste in the reactor and you get out the syngas. That's it."

THE GARBAGE MAN

After his tour of duty in Korea, Longo put himself through night school at the Brooklyn Polytechnic Institute. In 1959, engineering degree in hand, he got a job at American Machine & Foundry (AMF)—the same company that today runs the world's largest chain of bowling alleys—designing hardened silos for nuclear intercontinental ballistic missiles, such as Titan and Minuteman. "There was never a time I can remember when I didn't want to be an engineer," he says.

For years, Longo tried to convince his bosses at AMF to go into the garbage business (as manager of new product development, he was charged with investigating growth areas). "I knew a lot about the industry, how backward it was," he says. The costs to collect and transport waste were climbing. He was sure there had to be a better way.

In 1967 Longo quit his job at AMF to start his own business, called International Dynetics. The name might not be familiar, but its product should: Longo designed and built the world's first industrial-size trash compactors. "If you live in a high-rise or apartment building and dump your trash down a chute," he says, "it's probably going into one of our compactors."

When Longo started his company, it was still easier and cheaper to just haul the loose trash to the dump. But gas prices climbed, inflation increased, and soon business boomed. In a few years, there were thousands of International Dynetics compactors operating around the world. The machines could crush the equivalent of five 30-gallon cans crammed with trash into a cube that was about the size of a small television. "Our purpose was to condense it so it would be easier and cost less to bring to a landfill," he says.

Then, in 1972, Longo read a paper in a science journal about fusion reactors. "The authors speculated that plasma might be used to destroy waste to the elemental level someday in the future," he recalls. "That was like a spear in the heart, because we had just got our patents out for our trash compactors, and these guys were already saying there's a prettier girl coming to town," he says. "It would make obsolete everything we were doing. I resisted looking at the technology for 10 years. But by 1984, it became obvious that plasma could do some serious work."

By then, the principal component of today's plasma gasification systems, the plasma torch, had become widespread in the metal-fabrication industry, where it is used as a cutting knife for slicing through slabs of steel. Most engineers at the time were focused on ways to improve plasma torches for manipulating metals. But Longo had trash on the brain—whole landfills of trash. He was intent on developing a system that used plasma to convert waste into energy on a large scale. So he jumped ship again. In 1988 Longo sold International Dynetics and founded Startech.

PLASMA TO THE PEOPLE

"People kept asking me, 'If this is so good, Longo, then why isn't everyone already using one?'" he says, referring to himself in the third person, a device he relies on frequently to emphasize his point. "We had the technical capability, but we didn't have a product yet. Just because we could do the trick didn't mean it was worth doing." Trucking garbage to dumps and landfills was still cheap. Environmental concerns weren't on the public radar the way they are today, and landfills and incinerators weren't yet widely seen as public menaces. "We outsourced the parts to build our first converter," Longo says. "When we told the manufacturers we were working with plasma, some of them thought it had something to do with blood and AIDS."

Longo describes the development curve as "relentless." He teamed up with another engineer who had experience in the waste industry and an interest in plasma technology. "We didn't have computers. We did everything on drafting boards. But I was aggressive. And the more we did, the more it compelled us to continue." It took almost a decade of R&D until they had a working prototype.

"I felt like St. Peter bringing the message out," Longo says of his first sales calls. In 1997 the U.S. Army became Startech's inaugural customer, buying a converter to dispose of chemical weapons at the Aberdeen Proving Ground in Maryland. A second reactor went to Japan for processing polychlorinated biphenyls, or PCBs, an industrial coolant and lubricant banned in the United States since 1977 ("really nasty stuff," Longo says).

Longo realized early on that what would make plasma gasification marketable was a machine that could handle anything. Some of the most noxious chemicals, he knew from his decades in the garbage industry, are found in the most mundane places, like household solid waste. Startech has an edge over some of its competitors because its converter doesn't have to be reconfigured for different materials, which means operators don't have to presort waste, a costly and time-consuming process. To achieve this adaptability, Startech converters crank the plasma arc up to an extremely high operating temperature: 30,000°F. Getting that temperature just right was one of Longo's key developmental challenges. "You can't rely on the customer to tell you what they put in," Longo says. "Sometimes they don't know, sometimes they lie, and sometimes they've thrown in live shotgun shells from a hunting trip. That's why it's imperative that the Plasma Converter can take in anything."

A video camera mounted near the top of the converter at the Bristol plant gives me a glimpse of the plasma arc doing its dirty work. At a computer station near the converter, Lynch taps a few commands into a keyboard, and a loud hiss fills the room, the sound of steam being released from behind a pressurized valve. "You can use that steam to heat your facility and neighboring buildings," he says proudly. Next to him is an LCD monitor with a live video feed from inside the reactor. A vivid magenta glow fills the screen as I watch the plasma torch vaporize a bucket of cell phones and soda cans. A hopper at the top of the vessel dumps another load into the plasma reactor, and seconds later, it vanishes too. "The idea," Lynch says, "is that regardless of what you put in the front end, what comes out will be clean and ready to use for whatever you want." I've watched him operate the converter for nearly an hour, and I'm still stunned to see no smoke, no flames, no ash, no pollution of any kind—all that's left is syngas, the fuel source, and the molten obsidian-like material.

CATCHING THE LITTER BUG

Low transportation costs, cheap land, weak environmental regulations—these factors help explain why it took plasma until now to catch on as an economically sensible strategy to dispose of waste. "The steep increase in energy prices over the past two years is what has made this technology viable," says Hilburn Hillestad, president of Geoplasma. His company, which touts the slogan "waste destruction at the speed of lightning with energy to share," is negotiating a deal with St. Lucie County, Florida, to erect a $425-million plasma gasification system near a local landfill. The plant in St. Lucie County will be large enough to devour all 2,000 tons of daily trash generated by the county *and* polish off an additional 1,000 tons a day from the old landfill. Of course, the technology, still unproven on a large scale, has its skeptics. "That obsidian-like slag contains toxic heavy metals and breaks down when exposed to water," claims Brad Van Guilder, a scientist at the Ecology Center in Ann Arbor, Michigan, which advocates for clean air and water. "Dump it in a landfill, and it could one day contaminate local groundwater." Others wonder about the cleanliness of the syngas. "In the cool-down phases, the components in the syngas could re-form into toxins," warns Monica Wilson, the international coordinator for the Global Alliance for Incinerator Alternatives in Berkeley, California. None of this seems to worry St. Lucie County's solid waste director, Leo Cordeiro. "We'll get all our garbage to disappear, and our landfill will be gone in 20 years," he tells me. The best part: Geoplasma is footing the entire bill. "We'll generate 160 megawatts a day from the garbage," Hillestad says, "but we'll consume only 40 megawatts to run the plant. We'll sell the net energy to the local power grid." Sales from excess electricity might allow Geoplasma to break even in 20 years.

In New York, Carmen Cognetta, an attorney with the city council's infrastructure division, is evaluating how plasma gasification could help offset some of the city's exorbitant waste costs. "All the landfalls around New York have closed, incinerators are banned, and we are trucking our trash to Virginia and Pennsylvania," he explains. "That is costing the city $400 million a year. We could put seven or eight of these converters in the city, and that would be enough." The syngas from the converters, Cognetta says, could be tapped for hydrogen gas to power buses or police cars. But the decision-making bureaucracy can be slow, and it is hamstrung by the politically well-connected waste-disposal industry. "Many landfill operators are used to getting a million dollars a month out of debris," says U.S. Energy's Paul Marazzo. "They don't want a converter to happen because they'll lose their revenue."

Meanwhile, Victor Sziky, the president of Sicmar International, an investment firm based in Panama, is working with the Panamanian government to set up at least 10 Startech systems there. "The garbage problem here is exploding in conjunction with growth," says Sziky, who lives in Panama City. "We have obsolete incinerators and landfills that are polluting groundwater and drinking water. We've had outbreaks of cholera and hepatitis A and B directly attributed to the waste in landfills. There are a lot of people in a small country, and there's no infrastructure to deal with it." The project will be capable of destroying 200 tons of trash a day at each location, enough to handle all the garbage for the municipalities involved—and, says Sziky, to produce up to 40 percent of their electrical demand.

Panama's syngas will probably be converted to hydrogen and sold to industrial suppliers. The current market for hydrogen is at least $50 billion worldwide, a figure that is expected to grow by 5 to 10 percent annually, according to the National Hydrogen Association, an industry and research consortium. Analysts at Fuji-Keizai USA, a market-research firm for emerging technologies, predict that the domestic market will hit $1.6 billion by 2010, up from $800 million in 2005. The Department of Energy's Patrick Davis says that when the long-awaited hydrogen-powered vehicles finally arrive, the demand for hydrogen will soar. But he also notes that, to have an effect on global warming, it's critical that hydrogen come from clean sources.

That's one more idea that's old news to Longo, who, as usual, is 10 steps ahead of the game, already embedded in a future where fossil fuels are artifacts of a bygone era. For the past several years, he has been developing the Starcell, a filtration mechanism that slaps onto the back end of his converter and quickly refines syngas into hydrogen. As he says, "We are the disruptive technology." Longo has been working in garbage for 40 years, making his fortune by literally scraping the bottom of the barrel. Which is, it turns out, the perfect vantage point for finding new ways to turn what to most of us is just garbage into arguably the most valuable thing in the world: clean energy.

David Glenn

The (Josh) Marshall Plan

Break news, connect the dots, stay small

To get to the newsroom of Talking Points Media in Lower Manhattan, you need to visit a pungent block of cut-flower wholesalers on Sixth Avenue, then climb a narrow stairway to an 800-square-foot suite that might once have been an accountant's office. This modest space is the home of a news organization that—among several other notches in its belt—played a pivotal role in bringing the story of the fired U.S. Attorneys to a boil. Not only were the major dailies slow to pick up on the controversy, but a Capitol Hill staffer says that the House Judiciary Committee itself would have missed the firings' significance if not for the barrage of reports from Talking Points. When Alberto Gonzales, Kyle Sampson, and Monica Goodling testified before Congress this spring, they may have had the reporters in this obscure Flower District building to thank for the honor.

And one reporter in particular: Joshua Micah Marshall, the 38-year-old founder and editor of TPM, who has grown the operation from a tiny center-left political blog that he began at the end of 2000. (Back then, referring to himself as the "founder" or "editor" of anything would have been an act of self-deprecating bloggy humor.) Today, Marshall presides over a staff of four reporters—one of whom also serves as deputy editor—three associate editors, and a small army of unpaid interns. Their work is posted on a quartet of interconnected sites: Talking Points Memo, as Marshall's original blog is known; TPM Café, a two-year-old site devoted to policy and culture debates; TPM Muckraker, a year-and-a-half-old project that trawls for political scandal; and TPM Election Central. In total, the sites draw roughly 400,000 page views on an average weekday.

Marshall often says that he is annoyed by "blog triumphalism," which he described in 2004 as "an unrestrained belief that blogs or similarly situated sites can and should revolutionize all politics and media." But with his restless institution building, he has made as good a case as anyone for blogging's journalistic merits. From the very early days of Talking Points Memo, he has (by accident or design) cultivated an intense relationship with a well-connected set of readers—lawyers, activists, policy wonks, and veterans of intelligence agencies. Those readers have offered an endless stream of tips, and they have occasionally been deployed en masse to plow through document dumps from the Department of Justice or to ask members of Congress to publicly clarify their positions on Social Security.

"I think within TPM lies the DNA of the future of journalism," says Justin Rood, a former TPM Muckraker reporter who now works for ABC News. "In terms of its relationship with its audience, its ability to advance stories incrementally and to give credit to other news organizations, and its ability to get the story to readers—it's been able to foster a real spirit of collaboration."

Rood's vision is plausible enough—but it seems equally possible that TPM will be remembered 50 years from now as a brief efflorescence, as something like *I.F. Stone's Weekly*. Many bloggers will surely follow Marshall's lead and attempt to do serious original reporting; and some large news organizations will surely become looser and "bloggier" in their presentation, turning to readers for tips, commentary, and research assistance. (If you want a sense of what the *Washington Post* will look like a decade from now, one reasonable place to start is The Fix, the political blog written by Chris Cillizza.) But it's far from clear how many of those new projects will develop the kind of reporter-reader chemistry and hard-nosed reporting that Marshall has cultivated.

Talking Points Memo can be as casual and digressive as any blog. Marshall occasionally posts pictures of his baby son or writes about finding old Bob Dylan footage on YouTube. But there is not much that is casual about the Talking Points newsroom. By 9:00 in the morning, almost every chair is occupied, and the place has the hushed intensity of an air traffic control tower. A pair of interns wearing fat headphones monitor three flat-screen televisions mounted along a wall. Two of them are tuned to MSNBC and CNN, which seem to be airing an endless loop of stories about Chris Benoit, the professional wrestler who killed his family and himself. The third is tuned to C-SPAN, which is about to broadcast a Senate Foreign Relations Committee hearing. If anything interesting—or interestingly false—gets said during that hearing, the interns can use TiVo to post a short video excerpt online, along with text commentary. On a good day, that process can take as little as 15 minutes.

Marshall, who commands a large desk in a secluded corner of the room, is a large-framed man with the pensive, slightly distracted air of an ambitious graduate student— more John Kenneth Galbraith than Seymour Hersh. He doesn't immediately seem like someone who would pester congressional underlings for documents or spend late nights sweating over his small business's balance sheets. But listen in on one of his daily conference calls with his reporters (two of whom are based in Washington), when Marshall displays his steely side, and his miniature news empire suddenly begins to seem less improbable.

On this early-summer day, the call touches on a number of TPM's recent hobbyhorses: the stalemate over whether White House officials will testify under oath about the U.S. attorney firings; the various Senate proposals to wind down the Iraq war; real-estate shenanigans involving Alaska's congressional delegation. There is also a more wonkish topic: whom to invite to participate in the following week's TPM Café "book club" on U.S. policy toward Iran.

Marshall's interventions during the call are typically brief but sharp: What is that source actually up to? How are these subpoenas likely to play out over the next three weeks? Even if you can't break any news today on that topic, please take a couple of hours and write a post that lays out the context for our readers. Marshall is the dominant person on the call: his baritone voice is less tentative than those of his reporters, and it would be an intimidating voice if it weren't leavened with a hint of amusement. Indeed, on his TPMTV videos—a daily feature that began in April—Marshall often flashes a certain cat-ate-the-canary grin even when he is describing great crimes of state.

Marshall's troops generally share that temperament. Across the room, an associate editor named Andrew Golis is nursing an iced coffee and supervising the production of a daily e-mail digest sent to roughly 10,000 readers. Like most of the Talking Points staff, Golis is more than a decade younger than Marshall. He graduated from Harvard in 2006; while he was there, he started a political blog of his own and spent a summer volunteering for Howard Dean. In conversation, Golis is one part earnest Rawlsian liberal and two parts cocky journalist, calmly waiting to pounce on whatever new falsehoods emanate from Washington this afternoon. He's working two screens at once, using a laptop to instant message with six colleagues and a desktop to lay out the e-mail digest.

Several feet away sits deputy editor Paul Kiel, a former *Harper's* intern who was hired in late 2005 as one of TPM Muckraker's first reporters. Kiel's desk faces Sixth Avenue, away from his colleagues, and as he quietly works the phone he seems to be willing himself to believe that he's alone in the room.

Today Kiel is tracking, among other things, new subpoenas that the House Judiciary Committee has issued to force testimony from White House officials about the U.S. attorney dismissals. At certain stages of this story, Kiel has broken news; he was the first to report that Senator Arlen Specter had introduced last-minute language into the 2006 reauthorization of the Patriot Act that allowed the White House to replace U.S. Attorneys for an indefinite period without congressional oversight. Kiel's posts today and tomorrow won't contain any such scoops but will be creatures of aggregation, with links to coverage in Salon, a statement from the House Judiciary Committee, and testimony from a Senate committee hearing.

That is the way Marshall likes his coverage. When asked whether he would rather have more staff resources devoted to original reporting, he says, "I think we've got our percentages down pretty well. I think it's key to our model that we don't draw a clear distinction" between original reporting and aggregation. Marshall favors such a mix because he wants his reporters to serve as the "narrators" of complex, slowly unfolding stories. "Sometimes that will mean walking our readers through what's being published elsewhere," he says. New articles in mainstream dailies often contain facts whose full implications aren't explored, Marshall says, "either because of space or editorial constraints or because the reporters themselves don't know the story well enough. They're often parachuted in to work on these topics for just a few weeks."

In mid-July, TPM broke the news of a suspicious land deal involving Alaska senator Lisa Murkowski, and that story's trajectory neatly illustrates the site's typical blend of reporting, aggregation, and commentary. Senator Murkowski, it seems, bought a piece of riverfront property in 2006 from Bob Penney, a real-estate developer and major player in Alaska politics. The sale was made at a mysteriously attractive price, well below the land's probable $300,000 market value, and Murkowski had failed to fully report the deal in one or two ethics filings.

Most of the Murkowski posts were written by Laura McGann, a young TPM Muckraker reporter who was hired in May, having previously covered bankruptcy courts for the Dow Jones wire service. (McGann says that she had never heard of Talking Points before reading an article about Marshall in the *Los Angeles Times* in March.) McGann's initial salvo contained its share of online bells and whistles— photos of the property that were e-mailed by a reader, a link to the Senate ethics manual—but her coverage was also notably sober. McGann quoted denials of wrongdoing from both Murkowski's spokesperson and from Penney, and she even ended the post, in classic wire-service fashion, with a nonpartisan sound bite from the much-quoted Norman Ornstein. It was not the crude hit-and-run that skeptics of political blogs sometimes say they fear.

Three days later, the *Anchorage Daily News* picked up the story, with a front-page article that credited "the national political Web log tpmmuckraker.com" in its second paragraph. The *Daily News* nailed down Murkowski's purchase price ($179,400), which McGann had been unable to do. (Real-estate transaction prices are not public records in Alaska.) In the same edition, the *Daily News* published an editorial denouncing the sale ("a disappointing turn of events for a senator who had until this point served Alaska well").

From this point forward, the coverage on TPM was mostly a matter of linking to and commenting on coverage from the *Daily News* and other outlets. McGann continued to do a bit of original reporting—for example, she called a county assessor's office to vet Penney's claim that he hadn't seen the property's most recent assessment, and she unearthed an audio clip of Penney testifying at a state hearing—but most of her effort went into aggregation. Her posts were centered around links to the *Daily News*'s coverage, and her tone became more conversational. She offered pieces of context, including a catalog of other members of Congress who've recently landed in trouble over real-estate deals.

Still more casual was Marshall's own commentary at the Talking Points Memo blog. He sarcastically reviewed Murkowski and Penney's explanations for the sale: "Imagine that, a politically-wired Alaska moneyman wants the state's junior senator to live next door to him. Who can question that?" He also sketched—in his most conversational, just-between-friends voice—"a series of very weird little details about Murkowski's disclosure reports" that McGann had encountered during her reporting. "From an editor's perspective, it was a bit hard to know how to treat this," he wrote. "You don't want to go too far out on a thin reed dealing with what could be mere errors in filling out the form." (Ten days after McGann's initial report, Murkowski announced that she would sell the land back to Penney.)

This odd admixture of reporter, columnist, tipster, and ombudsman—often wrapped into the same post—is central to TPM's identity. Marshall values original reporting, but chasing scoops is not his only priority. Even if he and his colleagues decided to abandon original reporting entirely, TPM would probably still retain almost all of its audience. Marshall believes his role is to bring his readers the best journalistic efforts on a particular topic, even when those efforts have appeared in other publications.

There is occasional muttering that TPM fails to fully credit the newspapers whose reporting it aggregates. But Dean Calbreath, a reporter at the *San Diego Union-Tribune,* says that Marshall has "always been meticulous about crediting" his newspaper's work. Calbreath and his colleagues have worked for two years on the interlocking scandals involving the now-jailed U.S. representative Randy "Duke" Cunningham and his defense-contractor friends. TPM has often commented on the *Union-Tribune*'s coverage of those stories, and Calbreath says that TPM's posts, even when they don't appear to break news, still push the story forward. The site "provides reporters with sources that might not be at the top of our radar screen," he says. "Being based in San Diego, I'm not a big reader of *The Hill,* for instance. But by reading TPM, I can have easy access to [*The Hill's*] pertinent articles. The commentary at TPM, meanwhile, poses important questions that we might not have thought of on our own."

Rood, of ABC News, says that he sometimes found TPM's aggregation itch personally frustrating when he was on staff. TPM's readership peaks in the late morning and midday—exactly when he felt a reporter should be on the phone with sources. But because of the readership pattern, it is during those hours that TPM reporters feel compelled to write new posts. "That's not a complaint," Rood says. "It's just something that we had to work through. We were inventing this as we went along."

Phrases like "inventing as we went along" come up often in conversations about Marshall. "Josh has been through so many self-made phases that no one could have predicted," says Daniel Rodgers, a professor of history who supervised Marshall's senior thesis at Princeton. Marshall arrived there in 1987 from Southern California, where his father taught marine biology. (Marshall's mother died in a car accident near their California home in 1981.)

In crafting his thesis, which concerned the nullification debate in Virginia in the early 19th century, Marshall "figured out how a historical argument works, and he figured out what sources he would need," Rodgers says. "That's not at all inevitable. Not every college senior who is excited about history makes that leap into effectively working with sources. If you like, there's the thread between his college work and what he's doing now—the interest in investigative reporting."

Next came graduate school in history at Brown, but Marshall decided within a few years that university life felt too cloistered and that he would rather write for magazines. (He finally did finish his dissertation in 2003, long after abandoning any thought of an academic career.) In the mid-1990s, he supported himself in part by designing Web sites for law firms; to promote that business, he published an online newsletter about internet law, which featured interviews with scholars like Larry Lessig. In 1997 and 1998, he spun off articles on internet free speech for the now-defunct online publication Feed and for the *American Prospect,* which was then based in Boston. Shortly thereafter, he was hired as an associate editor at the *Prospect.*

Marshall soon grew to regret that connection. He and the magazine's top editors, Robert Kuttner and Paul Starr, found themselves in a long and tormented series of ideological quarrels—ones "that would make very little sense to anyone outside the world of the *American Prospect,*" Marshall says. "In my own way, I really liked Clinton and Gore, and [Kuttner] didn't like either of them." On questions ranging from trade policy to Monica Lewinsky, Marshall was a few notches more sympathetic to the White House than were his left-liberal bosses. He also fought unsuccessfully for the magazine to be more clever with its Web site. "A lot of things that we do here now involve aggregation and editorial sifting," he says. "I remember that from very early in my time at the *Prospect,* I argued that the way to get a lot of traffic was to provide that service."

In 1999, Bill Moyers and what was then known as the Florence and John Schumann Foundation made a $5.5 million donation that allowed the *Prospect* to expand; as part of that process, Marshall moved south and became the magazine's Washington editor. But distance did not improve his relationship with his bosses. By mid-2000, he knew that he would soon leave.

Scott Stossel, a former colleague at the *Prospect* who is now the managing editor of the *Atlantic,* recalls Marshall as having a rich knowledge of political history and a gift for framing stories. When younger reporters were hatching new articles, Stossel says, they would turn to Marshall for advice on whom to interview and what to read. But Marshall didn't necessarily seem like someone who would be successful in a corporate environment; he sometimes had trouble with deadlines, working long but irregular hours in clothing that was rumpled even by the creaseless sartorial standards of the left-of-center press.

In November 2000, five months before he finally quit the *Prospect,* Marshall started writing Talking Points Memo, in rough imitation of the early political blogs written by Mickey Kaus and Andrew Sullivan, whose loose-limbed style he admired. "I really liked what seemed to me to be the freedom of expression of this genre of writing," Marshall says. "And, obviously, given the issues that I had with the *Prospect,* that appealed to me a lot."

Marshall had already struck up a friendly acquaintance with the contrarian Kaus, and Kaus added Talking Points Memo to his blogroll. "It was probably that link that took me from, say, two readers to a hundred readers," Marshall says. "After that point, it sort of grew organically."

The early weeks of Marshall's blog were, inevitably, devoted to the Florida election imbroglio. His voice was sometimes precious (he thankfully soon abandoned the habit of referring to *himself* as "Talking Points"), but there was no mistaking the blog for a stodgy liberal policy magazine. ("Did Chris Lehane really call Katherine Harris 'Commissar Harris'? Chris, I'm on your side, man, trust me. But that kind of talk really doesn't help matters.") On the blog's third day, sounding a theme that would be echoed in tens of thousands of left-wing blog entries to come, he denounced a "supercilious, plague-on-both-houses" *Washington Post* editorial about the election aftermath.

When he quit the *Prospect* in early 2001, Marshall intended to earn a living as a freelancer, using the blog as a loss leader to advertise his skills. He had no notion of earning any money directly from it. But the freelance market was tightening, and Marshall found himself stringing together assignments "for no money at all" from Slate, Salon, the *Washington Monthly,* and elsewhere. (He also briefly wrote a political column for the *New York Post.*) He began to have flashes of doubt about the blog, wondering, "Why, when I was really only marginally able to support myself, was I spending all of this time doing something that couldn't make any money?"

Then three things happened. First, the blog's readership spiked dramatically, from 8,000 to 20,000 page views a day, at the end of 2002, when Marshall publicized Trent Lott's implicitly prosegregation comments at a dinner in honor of Strom Thurmond. With the help of readers' tips, Marshall demonstrated that the Mississippi senator had a long record of similar talk. Many other blogs, including some on the right, piled on, and the episode ended with Lott's resignation as Senate majority leader. Second, in late 2002, Marshall began to receive tip-jar-style contributions from readers— nothing much, but it was an early inkling that his audience might support the site. Finally, one day in 2003, Marshall got a pitch from Henry Copeland, a former freelance correspondent in Eastern Europe who had developed a new technique for selling advertising on blogs.

"It took me several months before I finally agreed to try it," Marshall says. "We were trying to work out an initial price point. This was all so new. Should we charge $5 to reach our audience? $1,000? We set our price, and a couple of weeks later we sold our first ad." By the end of 2004, Blogads.com (as Copeland's service is known) was generating around $10,000 a month for Marshall. He could stop scrounging for assignments at Slate and the *Washington Monthly*.

A half dozen or so other political bloggers took advantage of their new Blogads.com windfalls to quit their day jobs. Marshall did much more: he decided to raise additional money from his readers to expand his site, giving birth to TPM Café and TPM Muckraker. "Josh keeps upping the ante," Copeland says. "He says, 'Give me a new set of cards; let's play it.' It would have been easy for him to just keep blogging like mad with a simple design. His expansion efforts have sucked up a lot of energy that might have gone into perfecting the core blog."

Marshall's first explicit call for reader contributions came in late 2003, when he successfully asked for support to cover his travel costs for a 10-day trip to New Hampshire during primary season. (That appeal netted $6,000 in 24 hours.) In early 2005, he passed the hat for a far larger amount, to support the launch of TPM Café. That appeal netted $40,000 and allowed Marshall to hire his first full-time colleague. Another fund drive later that year took in $80,000, which permitted the hiring of Rood and Kiel and the creation of TPM Muckraker. As recently as this past March, Marshall asked for money to support a further staff expansion. Marshall says that on three occasions, he has received donations of $1,000, but never anything larger; the vast majority of his readers' gifts, he says, are in the range of $20 to $50.

The theory, Marshall says, is that the "pledge drives" should support a substantial portion of the first year of a new hire's salary—but that beyond that first year, the employee's salary should be covered by expanded revenue from advertising. At this point, Marshall says, roughly a third of the site's normal monthly revenue comes from Copeland's Blog ads (which currently charges $10,000 for a "premium sidebar" ad at TPM); another third comes from banner advertising brokered by other companies (recent banners at TPM have pitched cell phone horoscope services and the film *National Lampoon's Dorm Daze 2*); and the final third comes from NextNewNetworks, a start-up Web-video firm that pays TPM to create short daily video segments.

"We've never had any investment capital behind us," Marshall says. "So we have to be profitable every month. It's all on a kind of cash-as-you-go basis." Larger print-media companies have occasionally approached Marshall about buying or investing in the site—"that's even happening now with a couple of places," he says—but those conversations usually break down when it becomes clear that the investors are really only interested in purchasing Marshall's individual services. "I've got half a dozen people whose livelihoods depend on me," he says. "At a minimum, everyone working here now would need to still have a job."

Such deals are sometimes tempting, at least in the abstract, Marshall says. "I'll be 40 in a couple of years, I've got a new kid, so obviously getting an amount of money that would give me some financial stability is appealing." But his wars at the *Prospect* taught him that he would really rather not have anyone looking over his shoulder. "To the extent that we can make this work independently, it's hard to see why we would give that up," he says.

Having shepherded the expansion of TPM and a major redesign that was rolled out across the four sites this summer, Marshall would now like to pause for breath. "I think our ideal staff size is maybe a little bit larger," he says, "but not much." During the last two years, he says, he has spent so much time on financial and administrative minutiae that he has had too little time for long-form writing. A visit to the archives bears that out. In 2003, when he hit his stride as a blogger, Marshall often wrote essayistic, 800-word posts about the Iraq war, many of which hold up well. But during 2005 his posts were much more staccato and were often tied to the immediate twists in the congressional fight over Social Security. Some days you could be forgiven for thinking that you'd wandered into an AARP campaign blog.

If TPM represents a future for journalism, it isn't necessarily obvious how it will be replicated. There is no just-add-water kit that either the *New York Times* or a 24-year-old Medill graduate could use to build a similar site of their own because Marshall's relationship with his readers has evolved slowly and organically. "Part of the reason that Josh has succeeded," says Jay Rosen, a professor of journalism at New York University and the author of the blog Pressthink, "is that he didn't come at this as a Web evangelist. He's actually an old-fashioned political reporter who happens to be very open to the possibilities of the Web." During his blog's nascent years, Marshall used those old-fashioned virtues to gain the trust of Capitol Hill sources and of his fellow political correspondents. Hendrick Hertzberg, a senior editor at the *New Yorker,* says that Marshall's commitment to a certain measure of shoe-leather reporting is one of his fundamental virtues. "Talking Points isn't just parasitic on the dying corpse of the newspaper industry, the way certain other sites are," he says. Hertzberg adds that "Marshall is in the line of the great light-bulb-over-the-head editors. He's like Briton Hadden or Henry Luce. He's created something new."

Even if its model isn't directly replicable, TPM surely offers glimpses of the future. Omnibus commentary sites like Pajamas Media and the Huffington Post can seem frantic and unfocused compared to TPM, but they are both edging toward doing more original reporting of their own. "There's an enormous cultural disconnect between bloggers and journalists," says Richard Miniter, a *Wall Street Journal* veteran who was recently brought on to serve as Washington editor of Pajamas Media. "But that's slowly breaking down." Miniter has patiently persuaded his blogger colleagues at Pajamas that it can sometimes be acceptable to use anonymous sources. (Miniter, a conservative, adds that Marshall "is often dead wrong. On the other hand, without Josh there are a lot of good stories that would slip by. He's got a good eye, and he's a good writer.")

But even as Pajamas and other as-yet-unheard-of sites begin to mimic the TPM blend of reportage, aggregation, and snark, it seems safe to say that there will never be anything quite like Josh Marshall and his crew. Not many news organizations have been created from scratch by unemployed specialists in colonial New England history. As Marshall says, it's probably foolish to believe that blogs "can and should revolutionize all politics and media." But if the White House's claims of executive privilege in the U.S. Attorney affair lead to a minor constitutional crisis, keep in mind this 800-square-foot hothouse in the Flower District.

About the Contributors

Emmy Award winner **Ted Allen** is host of the Food Network series *Food Detectives*. He's been a judge on every season of Food Network's *Iron Chef America* and Bravo's *Top Chef*. Ted was the food and wine specialist on the groundbreaking Bravo series *Queer Eye for the Straight Guy*, which had a 100-episode run. He is the author of the cookbook *The Food You Want to Eat: 100 Smart, Simple Recipes* (Clarkson-Potter) and cowrote the *New York Times* best seller *Queer Eye for the Straight Guy: The Fab Five's Guide to Looking Better, Cooking Better, Dressing Better, Behaving Better, and Living Better*. Since 1997, Ted has been a contributing editor to *Esquire* magazine. He was a finalist for a National Magazine Award for his *Esquire* feature on the little-known phenomenon of male breast cancer. Ted also writes regularly for such publications as *Bon Appétit* and Epicurious.com. Ted holds an MA in journalism from New York University and a BA in psychology from Purdue University. He lives in Brooklyn, New York.

Michael Behar is a freelance writer based in Boulder, Colorado, who covers adventure travel, the environment, and innovations in science. His work has appeared in several publications including *Outside, Wired, Men's Journal, Mother Jones, Popular Science, Discover,* and *Air & Space*. Formerly, he was a senior editor at *Wired* and the science editor for *National Geographic* magazine.

Caleb Crain lives in Brooklyn and writes about literature and history for the *New Yorker* and the *New York Review of Books*.

Julian Dibbell has, in the course of over a decade of writing and publishing, established himself as one of digital culture's most thoughtful and accessible observers. He is the author of two books about online worlds—*Play Money: Or How I Quit My Day Job and Made Millions Trading Virtual Loot* (Basic, 2006) and *My*

Tiny Life: Crime and Passion in a Virtual World (Henry Holt, 1999)—and has written essays and articles on hackers; computer viruses; online communities; encryption technologies; music pirates; and the heady cultural, political, and philosophical questions that tie these and other digital-age phenomena together. He lives in Chicago, Illinois.

Cory Doctorow is coauthor of the Boing Boing blog, as well as a journalist, internet activist, and science fiction writer.

David Glenn is a senior reporter at the *Chronicle of Higher Education.* He has written for *Dissent, Lingua Franca,* the *Nation,* and the *New York Times Book Review.*

Thomas Goetz is deputy editor at *Wired* magazine, where he writes about the confluence of technology, science, and medicine. His work tends to follow themes, and lately he's been exploring the power of big data and openness in science. He has also worked at the *Wall Street Journal,* the *Industry Standard,* and the *Village Voice.* He blogs at epidemix.org.

Charles Graeber is a National Magazine Award–nominated writer and contributing editor to both *Wired* magazine and *National Geographic Adventure.* In addition to this prestigious collection of tech writing, his stories have been selected for both *The Best American Crime Writing, 2008* and *Best Business Stories of 2001* collection (Vintage Press) and as "Notable Stories" in *The Best American Nature & Science Writing* and *The Best American Travel Writing* collections (Houghton Mifflin, 2004). He has written for *GQ,* the *New Yorker, New York Magazine, Vogue, Outside Magazine,* and numerous other rags.

Alex Hutchinson is a contributing editor at *Popular Mechanics* and writes a biweekly column on the science of exercise for the *Globe and Mail.* Before becoming a journalist, he worked as a postdoctoral physics researcher at a National Security Agency lab. His first book, *Big Ideas: 100 Modern Inventions That Have Changed Our Lives,* will be published by Sterling Publishing. Alex lives in Toronto.

Walter Kirn is a novelist, essayist, and critic who lives in Livingston, Montana. His works of fiction include *Up in the Air,*

Thumbsucker, and *Mission to America,* two of which have been turned into feature films. A frequent contributor to the *New York Times Book Review,* he has served as a contributing editor to *Time* magazine and as the literary editor of *GQ.* His nonfiction articles and essays have appeared in the *Atlantic,* the *New Yorker,* the *New York Times Magazine,* and *Esquire.* His next book, *Lost in the Meritocracy,* a scathing memoir of his years at Princeton and Oxford, first appeared in the *Atlantic* and will be published by Doubleday in expanded form in the spring of 2009. Kirn will be the nonfiction writer-in-residence at the University of Chicago in the fall of 2008.

Robin Mejia has written for *Wired, Science, Mother Jones,* the *Washington Post,* and the *Los Angeles Times* and has produced for CNN. The 2005 Livingston Award for Young Journalists honored her documentary *Reasonable Doubt,* which looked at how problems at crime labs can land innocent people in prison. She lives in Santa Cruz, California, with her husband and can be reached at mejia@nasw.org.

Emily Nussbaum is an editor-at-large at *New York Magazine.* She lives in New York and writes frequently about pop culture and technology.

Ben Paynter has written for *Wired, Details,* and *Outside,* among other publications. His work has also been featured in the *Best American Sports Writing* series. This story and another one about animal cloning inspired episodes of *Wired Science* for PBS. He lives in Kansas City, Missouri.

Jeffrey Rosen, a law professor at George Washington University and legal affairs editor of the *New Republic,* is a frequent contributor to the *New York Times Magazine.* He is the author most recently of *The Supreme Court: The Personalities and Rivalries That Defined America* (Times Books, 2007).

John Seabrook has been a staff writer at the *New Yorker* since 1993. He is the author of two books, *Deeper: My Two-Year Odyssey in Cyberspace* (Simon and Schuster, 1997) and *Nobrow: The Culture of Marketing, the Marketing of Culture* (Knopf, 2000). His new book, *Flash of Genius, and Other True Tales of In-*

vention, will be published by St. Martin's in September. *Flash of Genius,* a film based on the title story, appears in theaters in October.

Cass R. Sunstein, a professor of law and political science at the University of Chicago, is the author of many books, including *Republic 2.0,* published in 2007 by Princeton University Press.

Clive Thompson is a science and technology journalist. He is a contributing writer for the *New York Times* and *Wired* magazine and writes regularly for other publications. He was a former Knight Science Journalism Fellow at MIT, and for five years he has run the popular tech-culture blog www.collisiondetection.net.

Acknowledgments

Grateful acknowledgment is made to the following authors, publishers, and journals for permission to reprint previously published materials.

"Doctor Delicious" by Ted Allen. First published in *Popular Science*, October 2007. Reprinted with permission of the author.

"The Prophet of Garbage" by Michael Behar. First published in *Popular Science*, March 3, 2007. Reprinted with permission of the author.

"Twilight of the Books" by Caleb Crain. First published in the *New Yorker*, December 24, 2007. Reprinted in the print edition of *The Best of Technology Writing 2008* only. Reprinted with permission of the author.

"The Life of the Chinese Gold Farmer" by Julian Dibbell. First published in the *New York Times Magazine*, June 17, 2007. Reprinted with permission of the author.

"How Your Creepy Ex-Co-Workers Will Kill Facebook" by Cory Doctorow. First published in *InformationWeek*, November 26, 2007. Reprinted with permission of the author.

"The (Josh) Marshall Plan" by David Glenn. First published in the *Columbia Journalism Review*, September/October 2007. Reprinted with permission of the author.

"Your DNA Decoded" by Thomas Goetz. Copyright © 2008 Condé Nast Publications. All rights reserved. Originally published in *Wired*. Reprinted by permission.

"The Pedal-to-the-Metal, Totally Illegal, Cross-Country Sprint for Glory" by Charles Graeber. Originally published in *Wired*, October 16, 2007. Reprinted with permission of the author.

"Breaking D-Wave" by Alex Hutchinson. First published in the *Walrus*, September 2007. Reprinted with permission of the author.

"The Autumn of the Multitaskers" by Walter Kirn. First published in the *Atlantic Monthly*, November 2007. Reprinted with permission of the author.

"These Images Document an Atrocity" by Robin Mejia. First published in the *Washington Post*, June 10, 2007, as "These Satellite Images Document an Atrocity." Reprinted with permission of the author.

"Say Everything" by Emily Nussbaum. First published in *New York Magazine*, February 2007. Reprinted with permission of the author.

"The Meteor Farmer" by Ben Paynter. First published in *Wired Magazine*, January 2007. Reprinted with permission of the author.

"The Brain on the Stand" by Jeffrey Rosen. Copyright Jeffrey Rosen. This essay originally appeared in the *New York Times Magazine* and is reprinted by permission.

"Fragmentary Knowledge" by John Seabrook. First published in the *New Yorker*, May 14, 2007. Reprinted with permission of the author.

"The Polarization of Extremes" by Cass R. Sunstein. First published in the *Chronicle of Higher Education*, December 14, 2007. Reprinted with permission of the author.

Text design by Mary H. Sexton

Typesetting by Delmastype, Ann Arbor, Michigan

The text font is Granjon, which was designed in 1928
for Linotype by George Jones using Claude Garamond's
(1499–1561) late Texte (16 point) roman as his model.
It is named after the sixteenth-century French printer,
publisher, and lettercutter Robert Granjon (1513–89).
    —*Courtesy adobe.com and myfonts.com*